*Calculus Explained*

*by the same author*

USE AND ABUSE OF STATISTICS

THE FASCINATION OF NUMBERS

# *Calculus Explained*

W. J. Reichmann, F.S.S.

METHUEN & CO LTD
and
SCIENCE PAPERBACKS

First published by Methuen & Co Ltd
11 New Fetter Lane, London EC4 in 1964
Reprinted 1969
SBN 416 27180 4

First published as a Science Paperback in 1969
SBN 412 20710 9

Reproduced and Printed in Great Britain by
Redwood Press Limited, Trowbridge & London



Distributed in the U.S.A. by
Barnes & Noble Inc.

# CONTENTS


| | | |
|---|---|---|
| | PREFACE | *page* vii |
| 1 | What is the Calculus? | 1 |
| 2 | Functions and Continuous Variation | 10 |
| 3 | Limits and Infinity | 24 |
| 4 | Basic Differentiation | 36 |
| 5 | Simple Differentiation Methods | 49 |
| 6 | Pause for a Paradox | 66 |
| 7 | Maximum and Minimum Values | 72 |
| 8 | Differentials | 93 |
| 9 | Basic Integration | 100 |
| 10 | Substitution Methods | 123 |
| 11 | Exponents and Logarithms | 135 |
| 12 | Integration Methods | 153 |
| 13 | Trigonometric Functions | 167 |
| 14 | Hyperbolic Functions | 184 |
| 15 | Plane Areas | 191 |
| 16 | Volumes and Surface Areas | 206 |
| 17 | Interpretation of an Area | 221 |
| 18 | Mechanics and Integration | 230 |
| 19 | Statistics | 251 |
| 20 | Partial Differentiation and Integration | 260 |
| 21 | Curvature | 269 |
| 22 | Elementary Differential Equations | 279 |
| 23 | Expansion of Series | 296 |
| 24 | Fallacies | 305 |
| | APPENDIXES | |
| I | Partial Fractions | 315 |
| II | Trigonometric Ratios and Identities | 317 |
| III | Standard Integrals | 320 |

IV Logarithmic Equivalents of Inverse Hyperbolic Functions *page* 323
V Limits and Continuous Functions – Definitions 325
INDEX 327

# PREFACE

For many years the calculus was popularly supposed to be an almost insuperable barrier stretching right across the field of mathematics. To those who have not been initiated into the mysteries of the subject it still appears that to cross into the land of calculus requires an effort no less sustained than that required in breaking through the sound barrier.

Advanced calculus or analysis is of course no less difficult than any other subject at an advanced level and it would be foolish to suggest otherwise. But the basic operations of the calculus, with which we are concerned here, are not difficult at all. Anyone who has managed to handle algebra and to construct graphs is already equipped for a venture into calculus territory.

It is unfortunately true however that the calculus is often presented in such a way as to make it appear difficult – if not actually incomprehensible. The reader is too often plunged into a bewildering array of symbols without first having been given any idea of what the calculus really is and what it attempts to do. It is possible to teach the elementary rules in this way and indeed it has been suggested that it is better to do so than to try to explain abstract mathematical ideas. But if no attempt is made to explain the concepts on which the calculus is based simply because they are too difficult to explain, it is scarcely surprising if the reader finds the whole subject beyond him.

This book therefore sets out to describe the calculus, to explain its basic principles, its validity and its relation to practical problems, and to demonstrate the operation of its rules and methods.

Some aspects of the subject are puzzling on first acquaintance, either because they involve ideas to which we may be unaccustomed or because they appear to be based upon inconsistencies and paradoxes. Novel ideas there certainly are in abundance, but such difficulties as arise exist only in a misunderstanding of the calculus and not in the calculus itself. Experience of the kind of question which these difficulties have frequently prompted has made it very clear that an awareness of them is essential to any explanation of the subject and due allowance has been made for them here.

A great deal too much is also often taken for granted in the matter of the simplification and transforming of equations. As a result of this the reader who is trying to follow the principles of the calculus is apt to find his task that much more difficult, just as a motorist will be hampered in unfamiliar country if the signposts are removed. Replace the signposts and his task is that much easier again. That is what this book seeks to do.

*Chapter 1*

# WHAT IS THE CALCULUS?

The calculus is a most powerful branch of mathematics which, whether it is generally realized or not, enters into calculations having a bearing on almost every aspect of our moving world. Sir Isaac Newton was inspired to invent his particular version of it in order to explain the movements of the planets, and today it plays a prominent part in the planning of the orbits of space-ships and satellites.

Effective space research would indeed be impossible without the calculus and it is also true to say that most of the spectacular technological advances of recent years have to some extent depended upon its use. Its importance in nuclear physics may be gauged from the interesting fact that Newton's solution of the equation to describe the motion of the planets in the universe also describes the motion of an alpha particle in its bombardment of an atomic nucleus.

The usefulness of the calculus however is not restricted to the larger issues of natural physical phenomena. As in other branches of mathematics, a discipline evolved primarily for one specific purpose may usefully be applied to other purposes. In the social sciences, the calculus is prominent in the theory of statistics, in economics, and in life assurance. In general science it is important in the study of radio and electricity, optics, acoustics, thermodynamics, machine behaviour and design, and in engineering, mechanics and physics generally. At a more personal level the calculus enables us to calculate the most economical speed of running a car or to calculate the maximum volume of a Christmas present which may be despatched through the post or, perhaps, to determine the amount of ice-cream contained in a cone.

Calculus applications are indeed many and varied although normally many of these may be less obvious except to the specialist, since to understand the applications also requires an understanding of the technical field in which they are made. The calculus makes its most significant contribution in complex scientific problems but this obviously cannot be explained in its more advanced aspects unless the reader has an understanding of the principles of the branch of science involved. Nevertheless the basic principles of the calculus may be quite well demonstrated in less complicated subjects – such as consideration of car speed and acceleration or of the falling of a stone – and in these contexts the principles are easy to understand.

It is worth while noting here that many people happily employ formulae without realizing that the latter depend for their validity upon the processes of the calculus. Simpson's Rule is an example of this. Once we have learned that a particular formula is correct, we are able to employ it in specified conditions, just as we may eat a sausage without knowing how it was manufactured. Provided the formula is the right one and our calculations are accurate we may always obtain satisfactory results from its use, which is more than can always be said for the sausage! But a knowledge of the basic principles underlying the formula becomes essential immediately we are confronted with a problem which does not obviously fit the expected pattern.

The calculus provides a method or system of rules for calculating. What does it calculate? What type of problem does it help to solve? What is this tool that has placed such power in the hands of mankind? These are the questions we shall be discussing, but we must first clear some fundamentals out of the way.

The word 'calculus' is borrowed direct from the Latin and means simply a 'stone' or 'pebble'. Its connection with mathematics derives from the days when stones were used to assist the process of counting. Even today some cricket umpires count the number of balls delivered in an over by transferring a pebble for each ball from one hand to the other. By extension of meaning the word 'calculus' is further applied to wider fields of calculation even though these involve processes other than mere counting. Reference to *the* calculus however is intended specifically to the *infinitesimal calculus* or to one of its separate but complementary branches – *differential calculus* and *integral calculus*.

The infinitesimal calculus is so called because it was originally based upon the concept of numbers which are infinitesimally small. It is not a simple matter to explain in a sentence exactly what is meant by infinitesimally small, but at this stage it is sufficient to understand that it refers to smallness which cannot be distinguished by itself – smallness which if magnified many times would still be indistinguishable.

In many a continental cathedral will be found some statue, a part of which has been gradually worn away by the kisses of countless pilgrims through the ages. It might reasonably be thought, when a person kisses a stone, that his action could scarcely have any share in its erosion. That it does have such an effect becomes evident only after thousands of other people have also kissed the stone. The amount of stone worn away by a thousand kisses will not be discernible; how much less is the individual effect therefore of just a single kiss? Its effect is real but it is nevertheless so small that by itself it is of no significance at all.

The concept of the insignificance of an infinitesimally small number was an important one in the original ideas of the calculus although, as will be seen later, a more rigid formulation of the principles of the calculus could not be based upon the infinitesimal alone. Nevertheless an understanding of the infinitesimally small is necessary in the appreciation of the calculus and this subject is dealt with in Chapter 3.

In general the *differential calculus* provides a method for calculating the rate of change in the values of variable quantities – as for example it enables us to calculate the acceleration of a car (acceleration being the rate of change of speed) or the rate of growth of an organism. On the other hand the *integral calculus* provides methods for the calculation of dimensions such as areas and volumes of curvilinear shapes or solids, like circles or spheres, and, more important, for the measurement of the dimensions of mathematical curves generally.

These dimensions have different interpretations in a wide variety of problems and the integral calculus makes it possible to unravel some of the mysteries involved in these applications. It is for this reason that the integral calculus has been called the key to the universe, and in its directness of application it provides a much more powerful tool than does the differential calculus. This difference however is perhaps more apparent than real since the two branches of the calculus are inextricably interwoven.

The calculus has been described as a size language to measure motion but it is in fact much more than that. It has also been described as the algebra of change and this is certainly more apt since the calculus deals specifically with problems involving change, or *variation* as mathematicians prefer to call it. Motion is only one particular kind of change (that is, change of position) but it provides a good example of the need for the calculus.

The acceleration of a car at any instant is the rate of change of the car's speed at that instant. With the help of algebra we can calculate the *average* speed of the car over specified distances but algebra cannot assist in calculating the *exact* speed or acceleration at any instant of time.

Algebra and geometry are essentially concerned with problems which may loosely be described as static or which refer to objects which move at more or less uniform speeds. They cannot give accurate solutions to dynamic problems. Algebra does recognize the fact of change. In the equation

$$y = 2x^2$$

it recognizes that if the value of $x$ varies then so does the value of $y$ vary so that $y$ is always twice as great as $x^2$. But algebra alone is inadequate to deal with the measurement of the rates at which these quantities are varying.

Problems posed in algebra often make unreal assumptions. A train leaves London at 8 a.m. for a destination exactly 100 miles away and covers the distance in exactly two hours. If it maintains a uniform speed of 50 m.p.h. throughout the journey, at what time will it pass through a point Z just 10 miles short of its destination?

This sort of problem, it is true, is posed to test the student's ability in calculation rather than to gather any information concerning the train's progress, yet it contains within itself an indication of the limitations of algebra. Algebra can supply an answer only if the train maintains a uniform speed, but uniform speed throughout most of the journey is improbable because of the real conditions with which the train must contend. Furthermore, uniform speed throughout the *whole* journey is quite impossible. The train has to start and it has to stop. It has to accelerate and it has to decelerate.

The element of error involved in the answer to this particular problem resulting from the unreal assumption of uniform speed may be small. It may be so small as to make virtually no difference. Instead of passing through point Z at 1·8 hours after leaving London the train may pass through at 1·81 hours. This represents a difference of only 36 seconds. Whether an error of this size is important or not depends upon the circumstances and recalls the statistical maxim that an apparent difference is only a difference if it makes a difference! People who have to meet long-distance trains would be exceedingly happy if they were never delayed more than 36 seconds.

If however we want to know something about the train's progress during the moments of acceleration or deceleration, we could not expect any assistance from algebra. During these periods the speed is changing rapidly every instant.

Continuous change and rate of change is everywhere around us, some of it natural, some of it man-made, whether it be in the growth of a plant, in the speed of a car or of a falling bomb, or in the movement of the earth relative to the sun. Nature is essentially dynamic and rates of change are of extreme importance in a world where nothing remains identically the same for ever. Nature is always changing and we need the calculus to keep up with it.

Rates of change or variation are more particularly the concern of the differential calculus, while the integral calculus is more concerned with the dimensions of shapes. This statement of course merely provides a guide as to the main basic uses of the two branches of calculus. The two branches are in fact complementary and it will be seen later that the process of the integral calculus may be regarded as the inverse of the process of the differential calculus.

It may at first be thought that the measurement of acceleration and the measurement of shape involve such widely different concepts that there could be no connection between the methods employed. In fact the discovery of the calculus was itself greatly assisted by the introduction of co-ordinate geometry whereby variation in the values of variable quantities could be represented by a shape in the form of a mathematical curve or graph. Thus acceleration may itself be represented by a curve or it may be calculated from the properties of curves of related variable quantities such as speed, time, and distance travelled. Measuring the acceleration and measuring the dimensions of the curve representing the acceleration must give the same result.

It has already been noted that the calculus is not merely confined to problems involving what is normally understood to be described as motion. It is concerned with variation of every kind. Another connection between the concept of variation and the concept of shape may be demonstrated by the curvature or 'turning' of a road. The calculus can measure the rate at which a road turns or the rate at which a curved surface is 'moving' away from the horizontal. This is not far removed from intuitive ideas; one often speaks of a road turning quickly or sharply. Thus a basically dynamic method may be applied to certain problems which at first appear to involve only static considerations. The road does not 'move' in the same way as a car moves but its line of direction changes almost continuously along its length.

This connection may be more clearly understood if we imagine a car to be following the winding of the road. The car is continually changing its direction because the road does so. If we wish to measure the rate at which the car is turning at any instant in its progress (that is, rate of turn, not the forward driven speed) this is clearly the same as the rate at which the road turns, and both of these rates, being identical in value, may obviously be measured by one and the same process.

There are clues in all this as to the origin of the calculus. Mention has been made of lines of direction and curves – and the calculus has a great deal to do with the properties of curves. The distances travelled by an accelerating car after certain specified periods of time may be plotted on a curve. The direction of a road may be traced out on squared paper so as to give the semblance of a curve, although this of course will not be a true mathematical curve. A study of these outlines, which are geometrical representations of particular forms of change, enables us to calculate the rate at which the change is taking place.

But although geometrical constructions help towards an understanding of the calculus, once we have this understanding we may extend our knowledge to similar sets of circumstances without always having to refer to geometrical reconstructions of the problems involved. In geometry, for example, once we have proved Pythagoras' Theorem for ourselves, we may apply the theorem to every right-angled triangle without having to prove the theorem for each triangle.

Furthermore when we have noted the relationship between certain classes of numbers in the equation ($a^2+b^2 = c^2$) and have started to ask why such a relationship should exist, we are led on to find other methods for calculating possible values of $a$, $b$, and $c$ – methods which involve purely arithmetical processes and which do not rely on the properties of a right-angled triangle at all.

Similarly in the calculus we do not always need to draw our curves even though we may often be aware of their underlying presence. Instead we are able to set up equations in a special notation and to work direct from these. Leaving behind constructional representation the mathematician is free to consider related problems which cannot be represented in a simple geometrical construction.

The calculus was first invented as a method of solving problems associated with physical phenomena. Once this basis was formulated however it was possible to develop the theory along analytical lines within the mathematical framework of the subject without worrying about the physical problems to which the mathematical abstractions might be applied.

This is often puzzling to beginners in any branch of mathematics and is particularly so in the calculus where the notation is at first especially strange. Yet the whole body of mathematics depends upon the abstract approach. Geometrical propositions, for example, assume that a line has no thickness. The mathematical idea here is abstract but it may be applied successfully to physical problems. It is indeed true to say that although mathematics may be rooted in experience, nevertheless much of its development is achieved away from physical applications and within the confines of mathematics itself.

But abstract ideas cannot be handled mathematically unless there is a suitable notation in which to express them. For the notation of the calculus we are indebted to Gottfried von Leibniz, the German philosopher and mathematician. Laplace referred to this as a 'happy notation', thus expressing the mathematician's gratitude for what was virtually a new language. Laplace, however, was himself a famous mathematician – newcomers to the calculus are generally less enthusiastic and it is paradoxical that the very notation which was invented to make life easier for the mathematician itself contributes largely to the mystification of others.

Let the reader be reassured. The basic calculus notation is no more difficult to learn than is the notation of algebra. Here the notation will be introduced only as it is required for the development of the theory. The theory may best be introduced by the use of the notation, and the notation can best be explained in its relation to the theory.

No book on the calculus is ever regarded as complete unless it refers to the controversy surrounding the origins of the subject. A great stir was caused in the seventeenth century over the vexed question of whether Leibniz or Sir Isaac Newton had first discovered the calculus. Nearly simultaneous discoveries in mathematics are not uncommon and, viewing the controversy from afar, it now seems probable that Newton and Leibniz made their discoveries of the general principles quite independently of each other and at about the same time. There can be no certainty of this. It was characteristic of Newton's work that he studied for himself. Publication of his findings was not of prime importance to him and there is really no way of discovering when he actually conceived his theory.

While it is thus impossible to resolve this relatively unimportant conundrum, it is nevertheless possible to discern different approaches in the work and thought of these two great mathematicians. If it were possible to single out the main contributions of each it would probably be fair to say that Leibniz had more to do with the fashioning of the calculus as a mathematical tool, while Newton's greatest contribution was undoubtedly to show how that tool could be used to interpret the observed behaviour of natural phenomena.

The controversy should not however be permitted to confuse facts by giving the impression that one or other of these famous men was fully responsible for all the discoveries made. Nothing would be further from the truth than to suggest this. No one individual is entitled to all the credit. The man who makes a break-through in science is dependent upon those who have preceded him in the same way as the mountaineer who first scales a peak depends upon the fellow members of his expedition and upon the experiences of previous expeditions. While it is his honour to be the first in the line of achievement and while his effort is not to be belittled in any way, he could never have got there alone. In his *History of Mathematics*, W. W. R. Ball has commented: 'It was Newton's good luck to come at a time when everything was ripe for the discovery, and his ability enabled him to construct almost at once a complete calculus.'

The calculus was developing over a long period of time before it was recognized for what it was. Mathematicians and philosophers had for years been groping – not always in the right direction perhaps – towards the principles which eventually materialized as the calculus. One of the seeds was in fact sown in antiquity by Archimedes. The introduction of co-ordinate geometry by Descartes and Fermat and the discovery by Isaac Barrow, Newton's teacher, of the properties of a particular triangle* were two essential steps towards the discovery of the calculus. Many other mathematicians had also made valuable contributions. What Leibniz and Newton did was to find the cement which could bind together all the previous discoveries into one body of knowledge to act as a foundation for their own discoveries.

Newton was one of the first to acknowledge his indebtedness to others. 'If I have seen further than Descartes,' he said, 'it is by standing on the shoulders of giants.' No man could have said more.

* Barrow's triangle. See page 40.

## *Chapter 2*

# FUNCTIONS AND CONTINUOUS VARIATION

It is in the genius of great minds to make momentous discoveries in apparently simple experiments. Galileo, the most famous son of Pisa, is said to have deduced his law of falling bodies from the simple experiment of dropping stones from different levels of the Leaning Tower. He first noted that in one second of time a stone fell a distance of almost exactly 16 feet – although of course Galileo used different units of measurement. Until then contemporary thought would probably have expected that a stone would therefore fall 32 feet in 2 seconds, that is, twice the distance in twice the time. In fact Galileo found that it took much less time than this for a stone to fall 32 feet.

Here was a problem. The relationship between time and distance fallen was not constant; in what way did it vary? We know now that a falling stone will travel the first 32 feet in approximately 1·414 (that is, $\sqrt{2}$) seconds, but Galileo's pendulum could not record such a measurement of time. He could measure only complete seconds. He therefore continued to climb the tower, throwing stones as he went, and he eventually found that a stone would fall 64 feet in the first 2 seconds of its fall. Further experiments showed that in 3 seconds a stone would fall 144 feet.

From these results it appeared that the relation between time and distance fallen could be expressed in the form

$$d = 16t^2$$

(where $d$ = distance in feet and $t$ = time in seconds). Thus for 1, 2, and 3 seconds respectively, the distance fallen was:

| $t$ | $d$ |
|---|---|
| 1 | $16 \times 1^2 =$ 16 feet |
| 2 | $16 \times 2^2 =$ 64 feet |
| 3 | $16 \times 3^2 =$ 144 feet |

As the stone made its downward fall its speed was actually increasing.

Galileo could not continue his experiments to test this relationship for 4 seconds. The Leaning Tower is only about 180 feet high. From the law deduced, the stone would presumably have fallen 256 ($= 16 \times 4^2$) feet in 4 seconds, and Galileo could not climb so high. How grateful he would have been for the modern skyscraper!

The importance of this discovery cannot be minimized but, as is often the case, the solution of one problem revealed other problems. The speed of the stone was clearly increasing during its fall. Its average speed over any specified distance could perhaps be calculated, but how was the speed at any specific instant of time to be determined? How was it possible to measure the acceleration (that is, the rate of increase of speed)? The mathematics of Galileo's days could not give an answer but they are typical calculus problems which nowadays may be solved quite simply.

The importance of being able to calculate the speed at which a body is travelling may easily be demonstrated. If you jump from a box which is only one foot from the ground, you will land on the ground in a fraction of a second and probably be none the worse for the jump. If however you jump from an aeroplane at 10,000 feet without a parachute you will certainly be killed. It is not merely contact with the ground which is injurious; it is the force with which contact is made that has the serious effect, and this force or impact would depend upon the speed at which you would be travelling when you hit the ground.

## CONTINUOUS VARIATION

The falling stone provides an excellent example of a use for the calculus, but before attempting to use this method it is necessary to consider some of the basic concepts involved.

In describing the progress of the stone reference has been made to its change of speed. In mathematics *change* or *variation* refers specifically to changes in the value of a variable quantity (which is simply a quantity which in differing circumstances may assume different values). Distance fallen and time elapsed are both variable quantities. They also possess the property of continuity.

A continuous variable is defined as being a variable the value of which, in passing from the value $a$ to the value $b$, assumes in succession all possible intermediate values between $a$ and $b$. This definition says much more than may appear at first. The expression 'all possible values' is to be taken literally. It includes all possible fractional and other values as well as whole numbers – and there is an infinitude of fractional numbers between any pair of consecutive whole numbers. There is also an infinitude of fractional numbers between any two fractions. Between $\frac{1}{4}$ and $\frac{1}{2}$ we may select another fraction $\frac{3}{8}$. Between $\frac{3}{8}$ and $\frac{1}{2}$ we may select $\frac{7}{16}$. No matter what two fractions we select and no matter how small these may be, there is always another fraction which lies halfway between them.

As we progress in this way the margins between the fractions become smaller and smaller but theoretically we may continue the process indefinitely, calculating the dimensions of a continuous variable throughout the whole interval during which it is varying. This is an example of an infinite process of which more will be said later.

The distance travelled by a motor-car travelling at uniform speed is a continuous variable dependent upon time elapsed. At 60 m.p.h. the car will cover one mile in one minute, a tenth of a mile in a tenth of a minute, a thousandth of a mile in a thousandth of a minute. If we so wish we may calculate the distance travelled during as small a fraction of a second as we care to consider. As we make the period of time progressively smaller, so the distance travelled during the period becomes smaller and smaller until the differences between apparently consecutive measurements become so small that they imperceptibly merge.

Temperature is also a continuous variable. It is measured in degrees, but changes in its measurement in fact pass through a continuous scale. If the temperature of a room is rising from 55 to 56 degrees then at some point in time the temperature must pass through a measurement of 55·00001 degrees although this could never be recorded. Changes in temperature occur quite independently of our ability to measure them. A change of temperature from 55·00001 to 55·00002 degrees would be so slight as to be imperceptible by itself. Nevertheless it is a real change and contributes its own small part of the total change from 55 to 56 degrees. This concept of the value of a variable varying at every instant is in fact the concept of continuous variation which assumes that, in passing from one value to another, the measurement of the variable passes through all possible intermediate values. Not all variables are continuous. A discontinuous or discrete variable can take only values which differ among themselves by certain fixed amounts. The number of goals scored at football is discontinuous; the number of goals in moving from 1 to 2 does not pass through 1·5 goals to any other intermediate values. It jumps direct from 1 to 2, for the simple reason that it is not possible to score a fraction of a goal. This jumping between values is the characteristic which distinguishes a discontinuous variable from a continuous one.

In the calculus we are concerned mainly with continuous variables although, as will be seen later, it is possible to adapt some of the principles to give reasonable approximations for discontinuous variables. It was the imperceptible merging of values which baffled earlier forms of mathematics in problems involving continuous variation. The value changes every instant and before one value can be identified and measured by static methods it will already have changed to another value, rather as a darting fish may already be down river before the fisherman can bring his rod into play.

The notion of change was but dimly grasped by the Greeks and their mathematical thought was conditioned by their general conception of the physical universe. Reality was everything, and it was traditional to regard reality as essentially stable. Yet there were some philosophers straining towards an understanding of continuous change. Heracleitus contended that nothing is stable, everything is in a perpetual state of flux. 'It is not possible,' he said, 'to step into the same river twice, because the second time it is not the same river.' This viewpoint was subsequently extended to point to the impossibility of 'stepping into the same river *once* since it is changing as we step'. It is the same, yet not the same. This goes far deeper than mathematics. It goes to the very root of existence, for how may we say that a thing is (that is, has an existence of its own) if it is always changing into something else? But the theory in this, even if it was understood and accepted, had little effect on the mathematical thought of the day. The universe might be split into the knowable (that is to say reality), and the unknowable; mathematics was still regarded as being concerned only with the former.

The position today is very different and the idea of continuous variation is one of the basic concepts upon which the calculus is based. It can handle these ideas quite easily and indeed it relies heavily upon them. There are some difficulties of interpretation, but these are based upon a misunderstanding of the principles and not on any inherent defect of the calculus and we shall not refer to them until they arise naturally in the development of the subject.

## FUNCTIONS

Another basic concept is that of *function* or the relationship between variables. It often happens that the variation in the value of one variable is always associated with the variation in the value of another variable. In the equation for the distance fallen by a stone

$$d = 16t^2$$

the variation in $d$ is always associated with the variation in $t$. As $t$ increases so also does $d$ increase, although not at the same rate.

Where variables are associated in this way, the variable whose value determines the value of the other variable is called the *independent variable*. The variable whose value is thus determined is called the *dependent variable*. This relationship is referred to as a functional relationship and the dependent variable is said to be a function of the independent variable. In simple language this merely means that the dependent variable is expressed in terms of the independent variable, so that when a value is assigned to the latter, there is thus determined a corresponding value of the former.

Use of the words 'dependent' and 'independent' does not imply any cause and effect relationship. It is usual to think of the area of a circle as depending upon the length of the radius because the radius is a basic requirement for drawing the circle. The radius and area however are but two of the many interdependent dimensions of a circle; neither may be said to be the cause of the other. If the length of the radius is varied then the area will also vary. Similarly if the area is varied, then the length of the radius must also vary. Dependence merely indicates that if we know the value of one of the variables then we may determine from it the value of the other. It is often largely a matter of convenience which of the variables we should treat as the independent one.

In discussing the general properties of functions it is traditional to express the independent variable by the letter $x$ and the dependent variable by the letter $y$. Instead of writing

$$d = 16t^2$$

we may write

$$y = 16x^2$$

where $y$ is distance travelled and $x$ is time elapsed. The relationship between the variables remains identical, we have merely changed the symbols or names of the variables. The choice of symbol can in no way affect the properties of the variable it represents.

Yet in changing the symbol we have made a transition from the particular to the general. The symbols $d$ and $t$ intuitively suggest distance and time, the symbols $y$ and $x$ do not. In the general case we may treat the function without worrying about what $x$ and $y$ represent. For a particular practical problem the variables will of course be identifiable, for indeed the problem would otherwise be meaningless. But the calculus has a general application and is not confined to one problem.

The same may be said for any other branch of mathematics. If there are three boxes each containing six objects then altogether there are eighteen objects. We do not need to know what the objects are (apples, oranges, stones) before we can make the calculation. The essential requirement is that the problem shall be expressible in some form to which the process of calculation may be applied. The essence of the calculus is that a functional relationship between the variables shall be identifiable.

The function $y = 16x^2$ is an example of an *explicit function* in that it explicitly or directly states the equivalent of $y$ expressed in terms of $x$. The general form of an explicit function is

$$y = f(x)$$

which indicates that $y$ is a function of $x$. This general form is used extensively in calculus theory.

The relationship between $y$ and $x$ may however be shown in a form which does not explicitly state $y$ in terms of $x$. The equation

$$x^2+y^2 = a^2$$

is an example of an *implicit function*. It expresses a functional relationship between $y$ and $x$ but does not explicitly state what it is. The equation needs to be solved to produce the related explicit function

$$y = \pm\sqrt{a^2 - x^2}$$

Mention has already been made of continuous variables. Continuity is also a basic property of the functions with which we shall be dealing. A *continuous function* is one which may be described by an unbroken mathematical curve; that is, when to each value of $x$ between $a$ and $b$, there corresponds a *real* value of the function $y$. (A mathematical definition is given in Appendix V.) A function may be continuous within certain intervals yet discontinuous within other intervals. The function

$$y = \sqrt{x^3 - x}$$

behaves in this way. When $x = 1$, then $y = 0$. When $x = 2$, then $y = \sqrt{6}$. But for *any* value of $x$ between 0 and 1, there is no real value of $y$. Thus, when $x = 0{\cdot}4$, then $y = \sqrt{-0{\cdot}24}$, and this is not a real number (i.e. it involves the square root of a negative number).

There are some special kinds of continuous functions to which the calculus cannot be applied, but we shall be restricting our attention to those continuous functions to which it *can* be applied.

## GROWTH OF A FUNCTION

It is now necessary to observe something of the behaviour of a continuous function. As a simple example, let us assume that a racing car proceeds at the uniform speed of 120 m.p.h. over a certain distance. Then in every minute the car travels 2 miles, and the distance which the car will have travelled after any number of minutes is given by the function

$$y = 2x$$

where $y$ represents distance in miles and $x$ represents time in minutes.

This is a simple function. As $x$ grows, so will $y$ grow, although the amounts of their growth will not be identical. The problem now arises as to how we may calculate the rate at which $y$ is varying relative to the variation in $x$. In order to demonstrate this, let $x$ be increased very slightly by the addition of an increment represented by the symbol $\Delta x$, so that $x$ grows to $x + \Delta x$.

Before proceeding further it should be explained that $\Delta$ is the Greek capital letter *delta*; some textbooks use the small letter delta (thus $\delta$) instead of $\Delta$. In algebra the expression $ab$ means '$a$ multiplied by $b$'. But the symbol $\Delta x$ does not mean '$\Delta$ multiplied by $x$'. $\Delta x$ is to be treated as one symbol only, not as two associated symbols, and it means quite simply 'an increment' or, more crudely, 'a little bit of $x$'.

As $x$ increases, $y$ also increases. Let the resultant increment in $y$ be expressed by the symbol $\Delta y$. But the value of $y$ is always twice the value of $x$. Consequently the new value of $y$ (which has grown to $y+\Delta y$) must be twice the new value of $x$ (which has grown to $x+\Delta x$). Then

$$y+\Delta y = 2(x+\Delta x)$$

An illustration of this will be useful. Let $x$ and $y$ be represented by the base and height respectively of a right-angled triangle ABC (as in Fig. 1) so that the height is twice the base. Now suppose that the triangle 'expands', while retaining its original angles, to form the expanded triangle DEC. The base CB ($= x$) will have become CE ($= x+\Delta x$) and the height AB ($= y$) will have become DE ($= y+\Delta y$). But the height of the expanding triangle is always twice the base because the angles do not alter; the original triangle and all of its expanded forms are similar triangles.

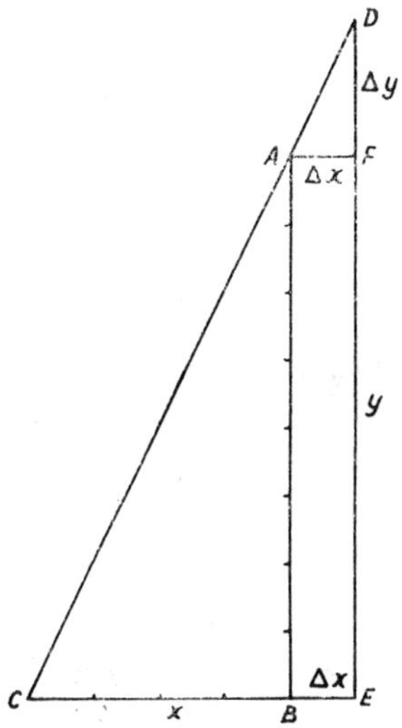

FIG. 1

This expanding triangle idea may be used to give us the value of $y$ for every value of $x$. Thus for the value $x = 4$ we arrange that the base shall measure 4 units (CB $= 4$) and then we can measure the value of $y$ along the perpendicular height of the triangle (AB $= 8$). This is in fact an illustration borrowed from co-ordinate geometry, the basic principle of which is that every equation or function has its curve and that every curve has its equation.

The car's progress may be shown as a curve (Fig. 2).

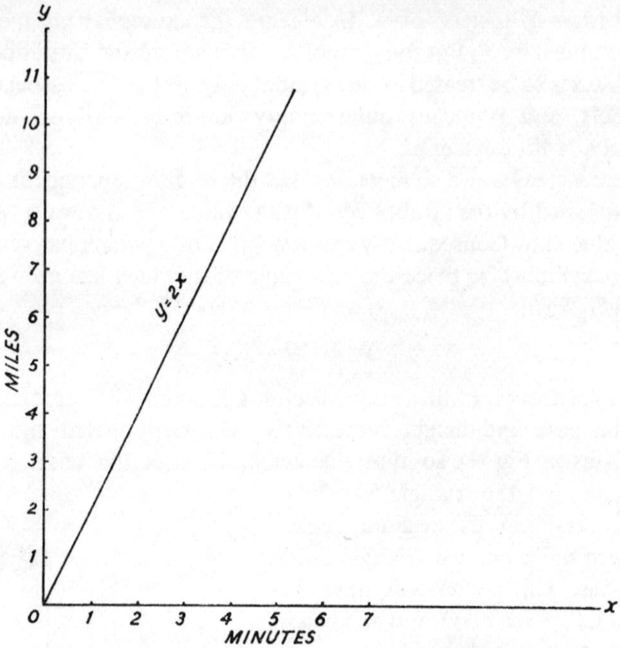


FIG. 2

Time is measured along the $x$ axis and distance is measured along the $y$ axis. The function $y$ is always twice the value of $x$ and the following co-ordinates may be plotted:

| $x$ | $y$ |
|---|---|
| 1 | 2 |
| 2 | 4 |
| 3 | 6 |
| 4 | 8 |

Joining the points thus plotted gives the linear (or straight-line) curve shown in the diagram. It will be seen that the hypotenuse of the triangle in Fig. 1 actually forms part of this curve.

It will be apparent that the slope of this curve relative to the axes is dependent upon the nature of the function represented. If we draw the curve of $y = x$, this will be less steep than our curve of $y = 2x$. The curve of $y = 3x$ however is steeper. This steepness of the slope of a curve is referred to as the *gradient* of the curve or of the function represented by the curve.

The idea of a gradient is well illustrated at approaches to dangerous hills where, for example, a road sign as in Fig. 3 may be erected (not drawn to scale).

FIG. 3

Strictly speaking, this sign should mean that for every 6 units of measurement which we travel horizontally we also travel 1 unit vertically, either uphill or downhill as the case may be, but in practice it is usual to replace the horizontal measurement by the actual distance travelled along the road. This gives an approximate result which is good enough for road signs but it does not hold good in mathematics. We assume here that the gradient has been calculated from the horizontal and vertical distances. This particular gradient is the gradient of the hill which is represented by the hypotenuse of the triangle, and it is represented as '1 in 6' or as the ratio $\frac{1}{6}$. This is a convenient way of expressing the gradient mathematically and is calculated from the triangle as the ratio

$$\frac{\text{height}}{\text{base}}$$

This is a ratio which we can recognize from trigonometry as being equivalent to tan $\theta$, the tangent of the angle $\theta$ between the hypotenuse and base of the triangle.

We now revert to the problem of the racing car as shown in Fig. 2. If on this curve we draw the $y$ ordinate at $x = 3$ we shall effectively produce the triangle in Fig. 4. The gradient of the curve between $x = 0$ and $x = 3$ may now be calculated in the same way as we calculated the gradient of the steep hill. At $x = 3$, the $y$ ordinate measures 6 units so that the ratio of height to base in this triangle is $\frac{6}{3}$ or, more simply, 2. This ratio measures the gradient of the curve and is therefore equal to tan $\theta$.

In Fig. 4 the height of the triangle represents distance and the base represents time. The ratio thus also expresses the speed of the car and we may summarize our results as:

$$\text{speed} = \text{gradient} = \tan\theta = 2 \quad \text{(in miles per minute)}$$

Again, if we draw the $y$ ordinate at $x = 2$, then the ratio of height to base of the new triangle thus obtained $= \frac{4}{2} = 2$. Similarly it will be found that, no matter where we draw our ordinate, the triangle formed will have a height-to-base ratio of $\frac{2}{1}$. This of course is because the speed is uniform throughout the curve. The angle $\theta$ is common to all such triangles. Since $\theta$ is constant then $\tan\theta$ (= speed) is also constant.

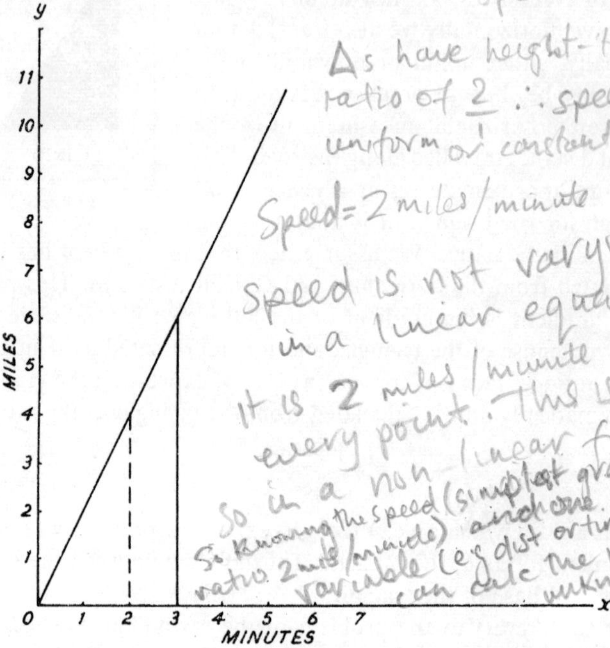


FIG. 4

For a more general method of tackling this kind of problem we revert to the idea of adding an increment to $x$ and seeing what happens to $y$. In Fig. 1, BE represents $\Delta x$, the increment added to $x$. AF is parallel to BE. AFEB is a rectangle. AF is therefore equal to BE (= $\Delta x$). DF is the increment added to $y$ as a result of the addition of $\Delta x$ to $x$. That is DF = $\Delta y$. In the triangle DFA we now have

$$\text{DF (height)} = \Delta y$$
$$\text{AF (base)} = \Delta x$$

But this triangle is similar to the triangle ABC. Consequently the height-to-base ratio is the same in both triangles. That is

$$\frac{\mathrm{DF}}{\mathrm{AF}} = \frac{\Delta y}{\Delta x} = \tan \theta$$

This is true no matter what value of $x$ we select as a starting point and no matter what size we make $\Delta x$.

In this illustration $\Delta x$ and $\Delta y$ have been drawn relatively large in order that it may be possible to identify them. But the properties of the triangle DFA when $\Delta x$ is large also hold good for similar triangles when $\Delta x$ is small – so small in fact as to be almost indistinguishable from a single point. Even though such triangles cannot actually be drawn on paper we know that the ratio of height to base remains identical at $\frac{2}{1}$. The speed of the car is uniform throughout its run. Although we knew at the beginning that this was so, we have now been able to deduce this from the properties of the curve.

This has been fairly simple in the present instance because the curve is a straight line. The gradient of the function is constant throughout the curve and in our calculation we have merely had to deal with a series of similar triangles. The properties of these triangles and their trigonometric ratios make our task an easy one.

The same principle applies to every linear function and in order to identify such functions it is necessary to define them in mathematical terms. A linear function of $x$ is one which does not involve any power of $x$ other than the first power. Thus the functions

$$y = 2x \qquad y = x+7$$

are both linear. On the other hand, the functions

$$y = 16x^2 \qquad y = x^3+2$$

are not linear. The curves of the second pair of functions are non-linear (i.e. curved).

The curve of $y = 16x^2$ is shown in Fig. 5. It will be recalled that this function represents the distance (in feet) fallen by a stone as a function of time elapsed (in seconds). The curve is non-linear throughout its length because the speed of the stone is changing all the time. Consequently the steepness of the curve is also changing. How then, if a curve is non-linear, can we find the speed at any instant? Since the speed is always varying, how can we trap a single value when it is not varying? It may seem self-contradictory to assert that we can do anything of the sort but, given a different interpretation of the problem in terms of infinite processes, we nevertheless can do just this with the aid of calculus.

For the linear function $y = 2x$ we could calculate the speed at any instant from the gradient of the curve. We could construct triangles for this purpose merely by drawing $y$ ordinates at the

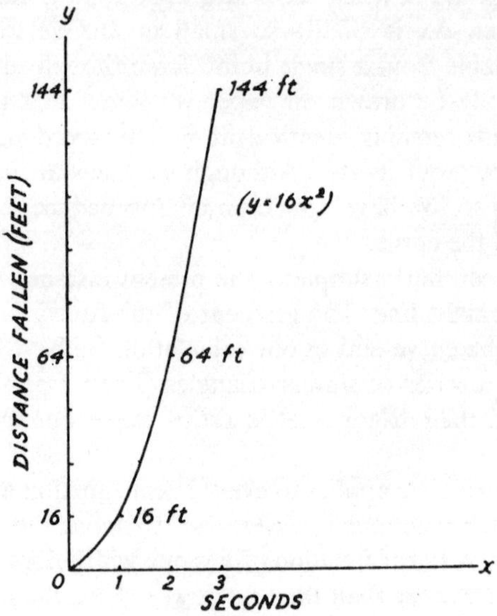


FIG. 5

relevant values of $x$. But the curve of the function $y = 16x^2$ is not a straight line and it is apparently impossible to construct triangles in the same way because a triangle cannot have one side which is curved.

The gradient of the linear function is equal to $\tan \theta$ where $\theta$ is the angle at the intersection of the curve and the $x$ axis (or any line parallel to the latter). We now have to use our imagination and think of a non-linear curve as being made up of a number of very small lines joined together and, indeed, made so small that within their own lengths they are practically straight.

Thus we may take the part of the curve between $x = 0$ and $x = 1$ and divide this into one thousand pieces relative to one thousand equal divisions of $x$, each such division being therefore equivalent to one-thousandth of a second. Then, although the *whole* curve is clearly non-linear, each small part of it is practically as straight as may make no difference. For each of these extremely small periods of time we may theoretically construct right-angled triangles so as to obtain an approximate idea of the speed by measuring the angle between the base and the hypotenuse.

This however will give only an approximation. In fact it will give the average speed over the length of the very small interval of time. It does not tell us the speed at any instant during that interval, although we can make an approximation as close as we like by making the time interval smaller and smaller.

The question now arises as to whether we can develop this idea any further so as to obtain an *exact* calculation of the speed at any instant. Furthermore, it is easy to talk about drawing very small triangles but in practice it is impossible to do so. A second question therefore arises as to whether we can develop the idea along purely arithmetic lines instead of having to rely on geometric constructions.

The short answer to both these questions is that we can do so, and that here we are at the very roots of the calculus. We may now state (without at the moment attempting to prove it*) that when a curve is non-linear, the gradient of the curve at any relevant point is equivalent to the gradient of the tangent to the curve at that point.

To understand this we must use the key provided by the calculus, but before we proceed to open the door we must first know something about limits and infinite processes. These deserve a chapter to themselves.

* See page 37 for proof.

## Chapter 3

# LIMITS AND INFINITY

The ancient Greeks regarded numbers rather as if the latter were controlling deities. Mathematics was for them a matter of dealing with straight lines or with simple curves of a uniform nature such as circles. They really believed in numbers for their own sake rather than for their application to the world around them. Such applications as they were prepared to venture merely seemed to them to confirm that the world, if not the universe, was constructed to a pattern of whole numbers.

In other material affairs where they themselves could see no way of applying mathematical principles, they concluded that mathematics could not be applied to such matters. Aristotle spoke of mathematics as being concerned with things which did not involve motion. Motion was regarded as a quality rather than a quantity and they had no way of applying quantitative ideas to such qualities.

The Greeks had no idea of the properties of the curve of a function involving non-uniform speed because the idea of co-ordinate geometry had not then been conceived. This method was made available by Fermat and Descartes only in the seventeenth century. The Greeks would have laughed heartily at the mere idea that motion could be represented by a curve. But even if they had known this to be possible they would still have been unable to deal adequately with the curve.

Nevertheless some of the Greeks were straining towards a further understanding of curves by the seemingly ingenuous method of pretending that the curves could in some way be treated as if they were composed of straight lines. This apparent paradox is more ingenious than ingenuous. While the methods they employed may seem unsophisticated to us who have at our disposal the accumulated theory of centuries of thought and research, this was a step out from what was at that time the accepted body of mathematical ideas. It must have seemed almost as heretical as Galileo's later assertion that the earth was round and not flat.

It may also have started the whole train of thought which was eventually to produce the revelation of the calculus, even though the Greeks themselves probably had no idea of where it would lead.

Their object was perhaps more to obtain approximate solutions to problems for which they could not calculate exact solutions. An example of this is given by the attempt to find a way of calculating the approximate value of the area of a circle. The philosopher Antiphon suggested that a polygon could be inscribed within a circle, the polygon having sides which 'in consequence of their smallness would coincide with the circumference of a circle'. This suggestion was too advanced for his contemporaries who pointed out that the inscribed polygon would in fact never coincide with the circumference of the circle, since to suppose that a rectilinear figure (that is, a figure having straight lines) could ever coincide exactly with a curvilinear figure (that is, having curved lines) was a contradiction in terms. The idea was unacceptable because it was opposed to their rigid conception of geometric principles.

Archimedes however used this idea of polygonal approximation in order to obtain a reasonably close value of $\pi$. If an equilateral triangle (the simplest form of regular polygon) is inscribed within a circle then the area of the triangle will be much less than that of the circle. If, now, another triangle is circumscribed about the circle, then this second triangle will have an area much greater than that of the circle. The same general relationship exists between a circle and other inscribed and circumscribed polygons but, as will be seen in Fig. 6, as the number of sides of the polygons is increased so do their areas approximate more closely to the area of the circle. The inscribed polygon becomes larger while the circumscribed polygon becomes smaller, thus 'trapping' the circle between them. When the polygons have a 'sufficient' number of sides, then these sides will be so small that the areas of the polygons will approximate sufficiently closely to the area of the circle.

Archimedes used this principle of increasing and decreasing polygons. He used polygons having ninety-six sides and by computing the dimensions of these he was able to show that the value of $\pi$ was less than $3\frac{1}{7}$ but greater than $3\frac{10}{71}$. Somewhere between these values was the true value of $\pi$ which was required to calculate the true area of the circle. The theoretical process involved in this method can be continued indefinitely with polygons having ever greater numbers of sides, but the calculations become progressively more cumbersome.

It will be noted that no attempt was made to say that this method did any more than give an approximate equivalence between a circle and a polygon. The polygon was never treated as being identically the same as the circle. If indeed the Greeks could have gone beyond this stage so that they could have bridged the gulf between the rectilinear and the curvilinear, then the principles

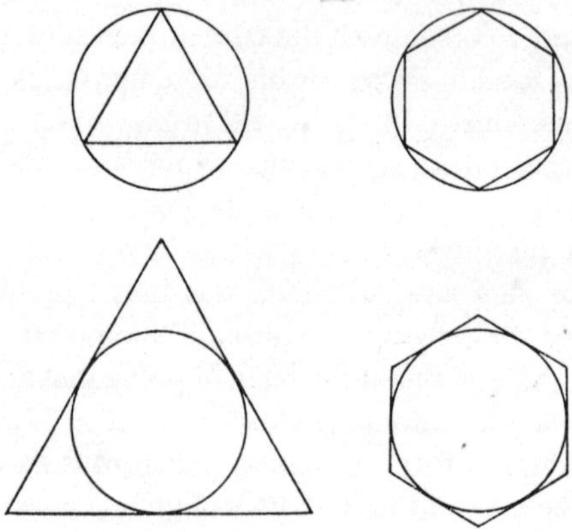

FIG. 6

of the calculus could have become known much earlier in the history of mathematics.

It was recognized that the process of drawing polygons could in theory be continued indefinitely or, expressed differently, that this was an infinite process wherein the polygons would approximate ever more closely to the circle without actually coinciding with it. This probably explains why the Greeks made no attempt to develop the method further and provides a clue to their rejection of the thoughts behind the very principles which today form the basis of the calculus.

The Greeks were happy with finite quantities but they could not cope with ideas of infinity. The infinitely great and the infinitely small were banned from their mathematics because they were concepts which did not allow of strict logical interpretation. The Euclidean view of mathematics excluded any idea which could not be clearly 'felt' through intuition. Aristotle denied the existence of the actual physical infinite and considered the term to represent a potentiality only. These and similar limitations were imposed upon mathematics; limitations which for many years were to confuse mathematical thought and to hinder mathematical development.

In the realm of pure number, the Greeks recognized the possibility of the infinitely great since it was obvious that, for any given number, a still greater number could be formed by adding 1 to the original number. This was in line with intuitive ideas and was therefore acceptable. But the Aristotelian view refused to admit the idea of the infinitely small because this was opposed to intuitive ideas. It is easier to think of something growing ever greater than it is to think of something getting smaller and smaller – as it shrinks, surely it will disappear altogether? Centuries later however Pascal was to point out that the infinitely great and the infinitely small are complementary. For every large number, such as 100,000, there also exists a small one which is the reciprocal of the other, such as $\frac{1}{100000}$. The mathematical existence of the infinitely great therefore implies also the mathematical existence of the infinitely small.

We have seen the Greeks 'trapping' the circle between inscribed and circumscribed polygons. The emphasis was thus in confining the circle within calculable limits. It apparently did not occur to them that, from the reverse viewpoint, the circle could be regarded as providing a limit to the polygons. The shape of the polygons progressed ever more nearly to that of the circle and the latter provided a shape beyond which the polygons could not progress. Yet a polygon with a finite number of sides could never be a circle, so that the theoretical construction of polygons having even greater numbers of sides was itself an infinite process. After we have drawn polygons having a million sides, the infinity of numbers implies that we can then draw polygons having a million and one sides. We cannot actually draw such polygons, but this is merely the result of our inability to represent certain mathematical truths; it does not invalidate the latter.

From thinking of an *infinite process* of constructing polygons with a *finite* number of sides, mathematicians came to think of a circle as being a *polygon with an infinite number of sides*. The area of any regular polygon is equal to the perimeter multiplied by half the perpendicular drawn from the centre of the polygon to one of its sides. If the circle is really a special form of polygon then its area should be calculable in the same way – and so it is. The perimeter of the circle is its circumference ($= 2\pi r$). The 'perpendicular' becomes the radius $r$, and the area is then

$$2\pi r \times \frac{r}{2} = \pi r^2$$

This step takes merely a sentence to express, but it represents a major change of attitude in mathematical thought which took centuries to develop. It involves ideas which are entirely opposed to intuition since it is impossible to think of a constructed polygon with an infinite number of sides. We must now study some of these ideas but first it should be noted that here we are talking of a mathematical infinity. We shall shortly note that the mathematical measurement of a stick of wood may in theory be divided into an infinite number of parts; it is nowhere suggested that the stick itself may be physically cut in the same way.

The infinite offers a challenge to thought and it is not easy to reconcile its concept with that of intuition. Intuitive ideas would prefer to recognize the number system as a closed and finite system rather like a fixed alphabet. The English alphabet has twenty-six letters while the integral number system employs only ten digits. But even so there is no finite limit to the number of combinations of letters or digits which may be derived if we do not specify any maximum number of letters or digits in any combination. There is nevertheless a very great difference between the structure of the alphabet and that of the integral number system.

The first five letters of the alphabet are A, B, C, D, and E, and it is not possible to envisage any letter between each of the consecutive pairs of letters. That is, there is no letter between A and B. The first five numbers of the integral number system are 1, 2, 3, 4, and 5. While there is no other whole number between any pair of consecutive numbers, it is possible (as was shown on page 12) to envisage other kinds of numbers between them. It is thus incorrect to describe the integral number system as comprising all possible numbers. As was previously noted there is an infinitude of numbers between any two consecutive integral numbers. This process of finding numbers between other numbers may be carried on indefinitely.

It is important to note however that if we are looking for numbers between 1 and 2, no number in which we are interested can *exceed* 2, since this is the limit which we have arbitrarily chosen. This idea of a limit is one which, in a particular aspect, must now be considered in some detail.

We have seen the circle as providing a limit to the construction of regular polygons even though that limit was never reached. This idea of *limit* is at the foundation of all our methods for dealing with infinite processes. In ordinary arithmetic, for instance, we can conceive of a series of numbers of the form

·9
·99
·999

and so on, each successive number being formed by placing another digit 9 at the end of the row of decimals. Each successive number is slightly greater than the preceding number, but the absolute difference between consecutive numbers in the series becomes smaller as the series progresses. The numbers are therefore growing at an ever slower rate, but this rate will never slow down so as to stop completely. The numbers in fact grow progressively closer to the integral number 1, but will never become exactly equal to that number. The number 1 thus provides what is called a *limiting value* to the series, this being the smallest number beyond which the series cannot proceed.

As another example we may consider the sequence

$$1 \quad \tfrac{1}{2} \quad \tfrac{1}{4} \quad \tfrac{1}{8} \quad \ldots$$

in which each successive number is exactly one half of the next previous number. The numbers in the sequence become successively smaller but will always be greater than nought – one half of a positive number must always be positive and thus greater than nought. The sequence will never terminate and is therefore infinite; there is no 'last' term. Here the limit is 0.

It is instructive to see what results may be achieved by adding the terms in this sequence. The first two terms sum to $1\tfrac{1}{2}$; the first three terms sum to $1\tfrac{3}{4}$. Summing terms in this way by adding a further term each time will provide successive totals as follows:

| | | |
|---|---|---|
| 1 | term | 1 |
| 2 | ,, | $1\frac{1}{2}$ |
| 3 | ,, | $1\frac{3}{4}$ |
| 4 | ,, | $1\frac{7}{8}$ |
| 5 | ,, | $1\frac{15}{16}$ |
| 6 | ,, | $1\frac{31}{32}$ |

For any given total in this set it is always possible to find another total greater than the given total but less than the value 2. There is in fact an infinitude of other possible totals between any total and the value 2. But as we add more and more terms of the sequence, the totals approach closer and closer to 2 and we can, by taking a sufficient numbers of terms, build up a total which is as close to 2 as we like to make it. We may therefore say:

(i) Given any total less than 2, we can always find another total closer to 2.

(ii) No finite number of terms can provide a total which is exactly equal to or greater than 2.

Clearly the number 2 provides a limit to the possible totals in much the same way as the circle provides a limit for the possible shapes of a polygon.

We have seen that no finite number of terms of the sequence may be added to make their sum exactly equal to 2. But if we accept that there is an infinity of terms, can we not also take the seemingly preposterous step of concluding that the total of this infinite number of terms is exactly equal to 2? How can we have a finite total of an infinite number of terms? This is seemingly self-contradictory, but to answer this question it will be found easier to reverse its process. We ask instead – can a finite total be divided into an infinite number of parts?

It was this question which was at the core of the paradox of the Dichotomy and other paradoxes with which the Greek philosopher Zeno plagued his contemporaries. How is one to build up an infinite number of parts to make a finite whole? The essence of Zeno's arguments is that anything which moves must always attain the middle point of its journey sooner than the end point of the journey, and it will take a finite time to reach this middle point. On the second half of its journey it must again reach the middle point of the second half of the journey before it reaches the end point. Each remaining part of the journey has a middle point which must always be reached before the end point; and since this process of dividing the journey may be carried on indefinitely the moving body can never complete the journey.

A practical example will be useful. A train travels from point A to point B exactly 100 miles away (see Fig. 7). The train must

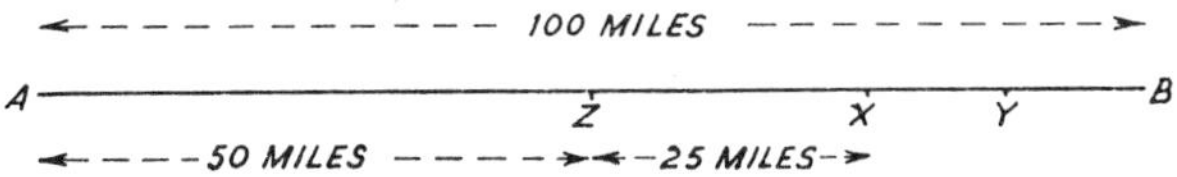


FIG. 7

reach Z, the mid-point, before it reaches B. From Z to B, the train must first reach X, the mid-point of the second half of the journey. Similarly, between X and B, it must first reach Y, the mid-point of the line XB. The same may be said of every part of the journey. No matter what distance the train has travelled it must always reach the mid-point of the remaining distance before it reaches the terminus. The distance may therefore be divided into an infinite number of finite parts. Each finite part requires a finite length of time, but the sum of an infinite number of finite intervals of time is itself apparently infinite. Accordingly the train will never reach its destination – there is always a little bit of the journey remaining!

This of course is not true, although we may sometimes impatiently believe it to be very probable when our trains are delayed. The distance is finite, for we have measured it so, and we know from experience that in the normal course of events the train will arrive at its destination eventually. How then is the paradox to be resolved? This has been a fertile field of thought for philosophers, but the point with which we are concerned here is that it is only possible for the problem to be posed because we can visualize that a distance may be divided into ever smaller pieces in an infinite process. We can agree that theoretically this process will never end, and this is tantamount to admitting that a finite length may be divided into an infinite number of parts. But intuition cannot easily accept the reverse of this theory – that an infinite number of parts may together form a finite whole.

Here again we have intuitive ideas in conflict with imaginative mathematical thought. But if we accept only intuitive ideas as a basis for mathematics, we may then admit only those ideas which are based on our own experience or on the accumulated and accepted experience of others. This would arrogantly suppose that we can confine the physical universe within the limits of our understanding of it. This is as absurd as saying that a tree is a tree only because we individually recognize it as such; the tree's existence is entirely independent of our own recognition of it. The world is greater than any of us, and an understanding of the physical laws to which the behaviour of the universe conforms depends upon the mathematics of the universe itself. A restricted form of mathematics, which rigorously insists that its scope shall be limited solely to what may be intuitively understood, is insufficient.

Zeno's arguments did much to expose the muddled thinking of his age and they have helped to clarify mathematical thought by forcing mathematicians to find the appropriate answers. They reveal a number of odd circumstances. Two of these are important. Firstly, although we can intuitively accept that a finite distance or similar geometrical type measurement may theoretically be divided into an infinite number of parts, we cannot easily accept that the measurement of time may be similarly divided. Secondly, although we accept the proposition that a finite whole may be divided into an infinite number of parts, we cannot easily accept the proposition that an infinite number of parts may have a finite total. We cannot justifiably say that the second proposition demands the use of ideas which are quite beyond our understanding of what is possible, for the same comment could then also be made about the first proposition.

While we can imagine that the division of a distance into parts can theoretically be continued indefinitely, we nevertheless know that in practical fact it cannot. This may be demonstrated by the cutting up of a wooden stick. Eventually, as we try to saw each remaining part into two pieces, we shall reach the stage when the piece of wood will be smaller in width than the width of the saw. In practice therefore the infinite division of the stick is not possible. Consequently in order to see the possibility of such a proposition, we have already had to resort to the use of mathematical thought as distinct from relying upon purely intuitive ideas. Merely because we can understand that the continuous division of remainders will always leave a smaller remainder, we delude ourselves into thinking that we can 'see' the process continuing indefinitely in fact as well.

The infinite series we are discussing exists only in the infinity of numbers, and this infinity exists only in mathematical thought. Nevertheless, although mathematical ideas do not always necessarily directly describe physical phenomena, they may still be applied to an understanding of them. You cannot multiply six apples by six apples to obtain the answer of six squared apples! This operation of multiplication also exists only in mathematics. Multiplication provides a short cut within the field of mathematics itself but its appropriate use can help to describe physical quantities.

Multiplication is possible only between numbers and not between the things which are numbered. In much the same way the mathematical length represented by the measurement of a length of stick may be divided into an infinite number of parts but the stick itself may not. The division of a finite mathematical length into an infinite number of parts is therefore not really an intuitive concept at all. If we accept such a proposition as a result of using mathematical ideas, it is not valid to reject the reverse proposition because it too requires the application of similar ideas when the real difficulty involved is merely that the effort of thought required is that much greater.

Zeno's arguments in fact contain their own solution for, if we suppose that a finite measurement may be divided into an infinite number of parts, then we must also accept that an infinite number of parts may together form a finite whole. The parts into which the measurement was divided may surely be reassembled to make the whole. The second proposition follows from the first, but the really important point to remember is that we are applying this concept of division to the mathematical measurements and not to the things or abstractions measured. The measurement of time may be just as easily dealt with in the same way as the measurement of length and consequently Zeno's argument that the sum of an infinite number of finite intervals of time must necessarily be infinite is seen to be invalid. We may relax in the knowledge that our train can reach its destination after all.

The subject of infinite series is a most fascinating one and a satisfactory exposition of it would require volumes. All that it is desired to establish here however is the basic fact that an infinite number of parts may be regarded as having a finite whole as their total. The finite total may be said to exist independently of the fact that we may divide it up into an infinite number of parts. Before this was recognized as a possibility it had been thought that, even if it were correct to think of a 'total' of an infinite number of parts at all (and this was doubted), then such a total must also necessarily be infinite.

Before proceeding further it is necessary to note that there are different kinds of infinite series – divergent series and convergent series. A convergent series is one the sum of the first *n* terms of which cannot numerically exceed some finite quantity however great *n* may be. Generally (but not always) it may be recognized in that the common ratio between successive terms is less than unity. Thus the series

$$1+\tfrac{1}{2}+\tfrac{1}{4}+\tfrac{1}{8}\ldots$$

is convergent since the common ratio is $\tfrac{1}{2}$. Each term of the series is multiplied by $\tfrac{1}{2}$ to obtain the next succeeding term. We have seen (page 29) that the terms of this series become smaller and smaller approaching as near to nought as we care to proceed. The series is thus approaching the value nought as we increase the number of terms and in this example the sum of *n* terms converges upon or approaches a finite total (which in this particular series is 2).

A *divergent* series however is one the sum of the first *n* terms of which can be made numerically greater than any finite quantity by taking *n* sufficiently great. Generally (but not always) it may be recognized in that the common ratio between successive terms is greater than 1. Thus the series:

$$1+2+4+8+16\ldots$$

is divergent since the common ratio is 2. In this series the terms become greater and greater and accordingly so does their sum become greater. This sum does not approach a finite total but instead is itself infinite.

Here we shall for the most part be concerned with convergent series in which the individual terms continuously approach zero and whose sum continuously approaches a finite total.

This total would be reached only if we could take an infinite number of terms; it cannot be reached by summing any finite number of terms. It is usual to speak of it therefore as a *limiting value* rather than as a total; that is, a value which applies to an infinite process and which we cannot therefore actually reach by finite arithmetic. The term *limiting value* or *limit* indeed has a wider connotation than the sum of an infinite series. The limit of an infinite sequence is the value to which the terms of the sequence continuously approach. The limit of the sequence

$$1 \quad \tfrac{1}{2} \quad \tfrac{1}{4} \quad \tfrac{1}{8} \quad \ldots$$

is therefore 0 and in this sense of course cannot be interpreted as a total.

A function $f(x)$ is said to tend to the limit $A$, as $x$ tends to the value $a$, if we can make the difference between $f(x)$ and $A$ as small as we like by taking $x$ near to $a$. (A mathematical definition of a limit is given in Appendix V.) This idea and its particular application to the calculus will be developed in the following chapter, but one general property of limiting values first needs special emphasis.

The existence of a limiting value to a convergent series implies that, although we have to deal with an infinite process which generally invalidates the use of finite arithmetic, yet we may associate with the series a value which is finite. We can never calculate the exact sum of the above sequence by adding its terms since the process of addition would itself be infinite. But we can show that the limiting value of this sum is 2. This value is finite and exact. If then the limiting value happens to be the value in which we are interested we are in the fortunate position of being able to derive an exact value as our answer. The sum of a finite number of terms is *approximately* equal to 2. The limiting value is *exactly* equal to 2.

## *Chapter 4*

# BASIC DIFFERENTIATION

With the limit concept at our disposal, we can now tackle the problem of the speed of an accelerating body. This, as was shown in Chapter 2, is a special form of the general problem of calculating the rate of change of a function.

For a linear curve it was found that we could calculate the rate of change at any point by measuring the gradient of the line at that point, and that the gradient was numerically equivalent to the tangent of the angle $\theta$ which the curve made with the $x$ axis. These ideas must now be applied to non-linear curves. When the curve is linear it is simple to measure the gradient at any point because in fact the gradient remains the same throughout the whole length of the curve. When the curve is non-linear however the gradient is changing throughout the curve and is different at every point. We now have to 'trap' the value of the gradient at a specified point.

Fig. 8 shows the curve of the simple function

$$y = x^2$$

for positive values of $x$. P is any point on the curve. What is the gradient of the curve at this point? The curve is non-linear. The rate of change is in fact increasing but, paradoxically perhaps, the best way to discover exactly *how* the rate is changing is to see what would happen if we were to suppose that it suddenly became uniform.

We therefore investigate what would happen if it be supposed that the rate of change were to increase right up to the point P but that thereafter it were to remain uniform at a rate equivalent to the rate at P. If the rate of change is uniform then the curve is linear. Hence, although this new curve would be non-linear as far as P, it would be linear from P onwards, in the direction of the arrow, to Z.

This line PZ is a very special one. It touches the original curve at only one point (that is, at P) and it is therefore the tangent to the original curve at that point. It is easy to see this. At every point on the original curve below P the rate of change is less than it is at P, since the rate is increasing at every point. For the same reason, at every point on the original curve above P the rate is greater than it is at P. No other point on the original curve can have the same rate as at P. On the other hand, the rate of change at every point on the line PZ is exactly the same as at P, for we have made it so. Consequently no point other than P can be on the original curve and also on line PZ. The line PZ touches the

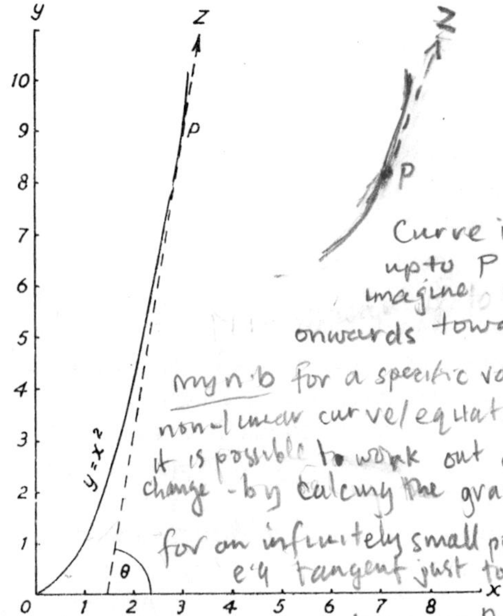


FIG. 8

curve at only one point and is, by definition, the tangent to the curve at that point.

But the rate of change at P is the same as it is throughout PZ. The rate of change along the line PZ is measured by its gradient – that is tan $\theta$. It follows that the rate of change at P is also tan $\theta$. P is *any* point along the curve and the principles which apply at P will also apply at any other point on the curve. Thus the gradient of the curve at any point is the same as the gradient of the tangent drawn to the curve at that point.

If, instead of thinking of a curve as being made up of points, we think of it as having been traced out by a single moving point, then what we have just discovered is another way of representing

Barrow's thinking of the tangent to a curve as demonstrating the direction of the motion of that point at the particular co-ordinate at which the tangent is drawn.

If we could draw the tangent to a curve at any point we would then be able to measure the angle $\theta$ and so to calculate the rate of change as tan $\theta$. For example, Fig. 8 has been so drawn that P is the point ($x = 3$; $y = 9$). If $x$ is time in hours and $y$ is number of miles travelled by a vehicle, then the speed (i.e. rate of change of distance relative to time) in miles per hour at P = tan $\theta$. A protractor will show that the angle $\theta$ is approximately 80° 30′. The value of tan $\theta$ may readily be found from tables to be 5·9758 (approx.). Consequently the speed of the vehicle represented at point P is approximately 6 m.p.h.

This result is shown as approximate because complete accuracy would depend upon our ability to draw the tangent correctly and then to measure $\theta$ exactly. It is easier to talk about drawing the tangents than it is to do the drawing. In fact, instead of drawing the tangent and then measuring the angle $\theta$, we need to know the angle first before we can draw the tangent! Where the curvature of a curve is almost imperceptible it is a practical impossibility even to begin the attempt to draw a line which will touch the curve at only one point.

In order to measure the angle $\theta$ therefore we cannot rely on constructions. But here the limit concept comes to our assistance. Fig. 9 again shows the curve of $y = x^2$; the scale of the drawing has been changed so as to make the details clearer. P and Q are separate points on the curve. PZ is again the tangent at P and, produced, meets the $x$ axis at A. QC is the $y$ ordinate at C. The points P and Q are joined by the straight line PQ; and the line PS, meeting QC at S, is parallel to the $x$ axis.

As a result of all this we have a most important right-angled triangle QSP, and it is on this that we now concentrate our attention. From the diagram it will be apparent that the gradient of the line PQ is not the same as the gradient of the tangent AZ. This is because the angle SPQ is greater than $\theta$.

If however we were to move the point Q along the curve towards P, the shrinking line PQ would move in a clockwise direction towards the line of the tangent and the angle SPQ would approach equality with the angle $\theta$. This is an infinite process. As Q moves nearer and nearer to P, so does the gradient of the line PQ approach more closely to the gradient of the tangent AZ. The tangent touches the curve only at P. But the line PQ touches the curve at two points (P and Q) and consequently, no matter how close Q is moved towards P without actually coinciding with it, the line PQ can never have exactly the same gradient as that of the tangent.

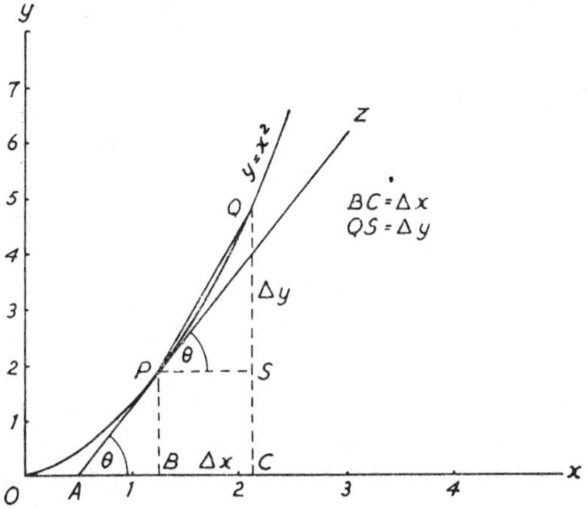


FIG. 9

We are thus faced with the proposition that we may make the gradient of the line PQ as nearly equivalent as we like to the gradient of the tangent without the two gradients ever becoming exactly identical. It was seen in Chapter 3 however that an infinite sequence may have a limiting value and that this value represents a real limit which can have an existence of its own independent of the infinite series.

Here we are concerned with two infinite sequences one of which is dependent upon the other. The first sequence is of the measurements of the line PQ as it continuously reduces. As the line becomes shorter its length continuously approaches the value zero. It never actually becomes zero, but no matter what measurement we care to specify (e.g. one-thousandth of an inch) we can always also specify a smaller measurement (e.g. one-millionth of an inch) which is closer to zero. Consequently the limiting value of this sequence is zero.

The second sequence is of the values of the angle SPQ as it approaches equality with the angle $\theta$. Here again, although we can specify a value very close to that of the angle $\theta$, we can always conceive of a value which is even closer to $\theta$. Consequently $\theta$ is the limiting value of the angle SPQ.

Thus, as Q approaches P, so the gradient of the line PQ approaches the gradient of the tangent. P provides the limit to the movement of Q, and the tangent provides the limit to the movement of the line PQ. It is the gradient of the tangent which we wish to measure in order to calculate the rate of change at P. To do this we first need to calculate the value of the angle $\theta$, and we now see that this may be done by treating the angle $\theta$ as the limiting value of the infinite sequence provided by measurements of the angle SPQ. Newton's teacher, Isaac Barrow, actually used this method for the construction of the tangent to a curve, and the shrinking triangle is sometimes referred to as *Barrow's Triangle* although it is better known as the *differential triangle* despite the fact that this is really a misnomer.

The proposition now needs to be formulated in a general manner. The argument is as follows, again referring to Fig. 9. P is any point on the curve such that OB $= x$ and PB $= y$. We now increase $x$ to $x+\Delta x$ by adding the increment $\Delta x$. As a result of this $y$ is increased to $y+\Delta y$. Thus we have:

$$\text{BC (and therefore also PS)} = \Delta x$$
$$\text{QS} = \Delta y$$

and in the right-angled triangle QSP (reproduced in Fig. 10) the tangent of the angle SPQ is equal to the ratio of the height QS to the base SP and is therefore equal to the ratio of $\Delta y$ to $\Delta x$. That is

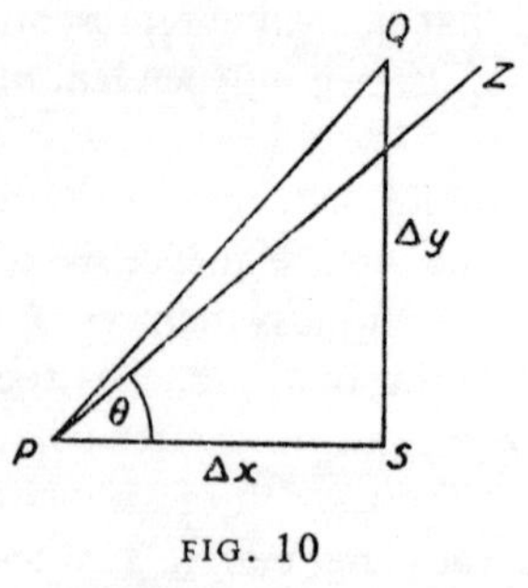

FIG. 10

$$\tan \widehat{SPQ} = \frac{QS}{SP} = \frac{\Delta y}{\Delta x}$$

This ratio expresses the gradient of the line PQ. As Q approaches P and the gradient of PQ approaches the gradient of the tangent to the curve, so the values of $\Delta x$ and $\Delta y$ are continuously reduced. $\Delta x$ approaches zero as its limit. As a result of this, tan SPQ approaches tan $\theta$ as its limit. But tan SPQ $= \frac{\Delta y}{\Delta x}$. Therefore $\frac{\Delta y}{\Delta x}$ approaches tan $\theta$ as its limit. It is the value of tan $\theta$ which we wish to calculate in order to find the rate of change at point P and this is now shown to be calculable as the limiting value of $\frac{\Delta y}{\Delta x}$.

This limiting value is usually shown as $\frac{dy}{dx}$ to indicate its association with $\frac{\Delta y}{\Delta x}$ and it is called the *differential coefficient* or the *derivative* of $y$ with respect to $x$. This is the quantity which the differential calculus has shaped as its tool. The use of the word *coefficient* is unfortunate and is misleading if confused with the use of the same word in algebra where it is used to denote a quantity which is one of the factors of a product. In order to avoid confusion the alternative *derivative* is used throughout this book. The process of finding the derivative is called *differentiating the function*; where the function $y$ is expressed in terms of $x$ (as in our examples) the process is called *differentiating y with respect to x*.

The relationship between $\frac{\Delta y}{\Delta x}$ and $\frac{dy}{dx}$ is formally expressed as

$$\lim_{\Delta x \to 0} \frac{\Delta y}{\Delta x} = \frac{dy}{dx}$$

This reads 'the limiting value of $\frac{\Delta y}{\Delta x}$, as $\Delta x$ approaches zero ($\rightarrow 0$) is $\frac{dy}{dx}$'.

It is sometimes said that $\frac{dy}{dx}$ is the limiting form of $\frac{\Delta y}{\Delta x}$ or that it is the 'ultimate ratio'. This however is not logically satisfactory. The limit would only be reached when $\Delta x = 0$. But then $\Delta y = 0$ also. In the limit therefore the ratio $\frac{\Delta y}{\Delta x}$ or $\frac{dy}{dx}$ as it would then be would be equal to $\frac{0}{0}$ and this is indeterminate. It will be necessary to deal with this question in more detail shortly but it is as well to be quite clear at once as to the nature of $\frac{dy}{dx}$. Although $\frac{dy}{dx}$ is expressed in the form of a fraction it should be regarded as one complete symbol representing the limiting value.*

The choice of symbol may perhaps seem a little unfortunate although in fact it will be found to be quite serviceable. It is sometimes unwieldy and an alternative notation is often used. We have seen that, if $y$ is a function of $x$, it may be expressed as

$$y = f(x)$$

Developing this idea, the limiting value of $\frac{\Delta y}{\Delta x}$ is sometimes shown as $f'(x)$.

This form of notation bears some resemblance to the notation introduced by Newton, but it is more meaningful. Instead of using the symbol $\frac{dy}{dx}$, Newton used the symbol $\dot{y}$ as the *fluxion*† of $y$, the dot indicating that the variable $y$ had been differentiated.

This notation imposed one serious restriction upon the understanding. Although it showed that $y$ had been differentiated it did not identify the independent variable in respect of which the differentiation had been effected. The symbol $\frac{dy}{dx}$ does not suffer from the same disability and, as will be seen later, it greatly facilitates understanding. It also proved to have inherent properties which were to assist in the development of the calculus.

It has been claimed that Newton's notation did much to delay the development of the calculus in England and more than a century after its introduction a society was formed at Cambridge 'to spread the principle of d-ism in opposition to the dot-age prevalent at the time'. The notation of $f'(x)$ avoids the ambiguity of $\dot{y}$ since it indicates that the independent variable is $x$.

* But see page 93 for the same symbol as the ratio between differentials.
† See page 69.

The symbols $\dfrac{dy}{dx}$ and $f'(x)$ mean exactly the same thing and we shall continue to use the former except in a few examples where its use becomes cumbersome.

As a first example of differentiation we now differentiate the function represented in Fig. 9:

$$y = x^2$$

Let $x$ (at point P) grow to $x+\Delta x$ (at point Q). As a result of this $y$ will grow to $y+\Delta y$. At every point on the curve the $y$ value is always equal to the square of the $x$ value. At point Q, $x$ has become $x+\Delta x$ while $y$ has become $y+\Delta y$.

Therefore $\quad y+\Delta y = (x+\Delta x)^2$
$\quad = x^2+2x\,\Delta x+(\Delta x)^2$

But $\quad y = x^2$

(Subtracting) $\quad \Delta y = 2x\,\Delta x+(\Delta x)^2$

(Dividing by $\Delta x$) $\quad \dfrac{\Delta y}{\Delta x} = 2x+\Delta x$

This equation holds for any non-zero value of $\Delta x$. As $\Delta x$ is made smaller and smaller approaching zero as its limit, so does $2x+\Delta x$ approach ever more closely to $2x$. This then is the limiting value of $\dfrac{\Delta y}{\Delta x}$ and it is therefore also the value of the derivative $\dfrac{dy}{dx}$. The derivative measures the rate of change and so we may say that the rate of change of the function ($y = x^2$) at any point $x$ is always equal to $2x$. How beautifully simple this final relationship proves to be and yet how much effort has been required to reach it!

The procedure we have adopted may be summarized: we start with the given function of $x$ which may be expressed as $f(x)$. In this instance $f(x) = x^2$. We add an increment $\Delta x$ to $x$ in the function which then becomes a function of $(x+\Delta x)$. That is, $f(x)$ becomes $f(x+\Delta x)$ and in our example $x^2$ becomes $(x+\Delta x)^2$. We subtract the original function from the new function, divide the difference by $\Delta x$ and then, to obtain the derivative, take the limiting value of the quotient. This is expressed mathematically as

$$\frac{dy}{dx} = \lim_{\Delta x \to 0} \frac{f(x+\Delta x)-f(x)}{\Delta x}$$

To check the result obtained we take the value of $x$ as 3. We originally tried to find the rate of change at this point by measuring the angle $\theta$ in Fig. 8 and it was found that, as near as we could measure it, the rate of change or speed was about 6 m.p.h. We now know that this result is exactly correct, for

$$\frac{dy}{dx} = 2x = 6 \quad \text{(in miles per hour)}$$

Similarly when $x = 4$, then the speed is 8 m.p.h.

As has been emphasized before, the value $2x$ will give only an approximate value for the general ratio $\frac{\Delta y}{\Delta x}$, the closeness of the approximation depending upon the size of $\Delta x$. But it is an exact value of the special quantity $\frac{dy}{dx}$ since this is the limiting value. This perhaps needs further clarification, as it is desirable to be very clear about what has just been said.

It will be noted, for example, that we have not claimed that our infinite series ever actually reach their limits. What we do claim however is that each of our series has a limit; that one series is dependent upon the other and that accordingly the limiting value of the one (that is, $2x+\Delta x$) depends upon the limiting value of the other (that is, $\Delta x$).

While the infinite series themselves will theoretically never reach their limits, nevertheless the properties of the series taken together enable us to calculate the limit of the dependent series. The two series are really used only to identify their limits and these, once identified, give us our exact results. Just because we employ an infinite series so as to identify its limit, it does not follow that we are bound to regard that limit as belonging only to the series. Thus, while the series has the apparent disadvantage that no finite number of terms can ever actually take us to its limit, this does not of itself deny the existence of the limit which may exist independently of our ability to reach it by any particular method.

We look instead for an alternative method. A cross-country runner may arrive at a river just after the only bridge across it has collapsed. He can see his destination – perhaps a church – on the other bank of the river but he cannot now reach it by running.

But because he cannot reach it, it does not mean that he cannot recognize it or that it does not exist. Its existence is real and we have to regard our limiting values in much the same way. If the runner can swim across the river or find a boat, he may still reach his destination despite the absence of a bridge.

For our part however we do not insist on performing any mathematical jumps. We are content to stay on one side of the void provided we can identify the limits on the other side. It is sufficient for our purposes that if we know the limiting value of one series, we may thereupon calculate the limiting value of the dependent series. We have set up an independent series of our own choosing by making $\Delta x$ approach zero. That is, we have selected a series of which we already know the limiting value. This is definite for we have so selected it. Thus in the limit, $\Delta x$ would be equal to zero. But if $\Delta x$ ever reached its limit, the dependent series of $2x+\Delta x$ would also reach its own limit; and at that point $\Delta x = 0$. Thus the limiting value of $2x+\Delta x = 2x+0 = 2x$ *exactly*.

It will be apparent that the concept of limits is of extreme importance since it is upon this that the theory of calculus is now respectably based. It has indeed made possible a mathematically rigorous formulation of the subject which was sadly lacking at the time of the calculus' introduction. At that time mathematicians were bemused by problems involving infinitesimally small quantities. They realized that they could safely ignore these but could not produce a satisfactory explanation of why this should be so.

The development of their theory was identical with that already demonstrated here up to the stage where they reached the equation

$$\frac{\Delta y}{\Delta x} = 2x+\Delta x$$

Here they said that when $\Delta x$ was infinitesimally small it could be ignored since it could be made as small as they liked. Nevertheless it had to be admitted that the infinitesimally small was still of some size and accordingly the convenient device of 'dropping' such quantities was opposed to strict mathematical principles.

This suggested that the resulting limiting value of $2x$ could only be approximate. Here was an apparent vacillation between calling an increment measurable at one stage of the calculations and then, at a later stage, treating it as if it were nothing. Yet the results were claimed as being exact, wherefore it seemed that a further paradox was involved in that mathematicians were using approximations to derive exact results.

Voltaire was moved to remark ironically that the calculus was the art of numbering and measuring exactly a thing whose existence could not be conceived. The most valuable criticism however came from Bishop George Berkeley:

> But it should seem that this reasoning is not fair or conclusive. For when it is said, let the increments vanish (i.e. let the increments be nothing or let there be no increments) the former supposition that the increments were something, or that they were increments, is destroyed; and yet a consequence of that supposition (i.e. an expression got by virtue thereof) is retained. Certainly when we suppose the increments to vanish, we must suppose their proportions, their expressions and everything else derived from the supposition of their existence, to vanish with them. . . .

Berkeley had no quarrel with the conclusions of the calculus. His quarrel was with the apparent illogicality of the methods employed. He complained that the calculus led from false premises to correct results because of the compensation of errors. Exactness of results and accuracy of method is essential to the basis of pure mathematics. If it is admitted that results are only approximate there is then raised a further question of how nearly accurate they are and whether the degree of accuracy is the same for every other similar problem.

Mathematicians are therefore particularly concerned to demonstrate that the methods of the calculus and the results derived from them are logical and exact, because of the lively criticism which has assailed the calculus from time to time. It is worth noting too that the approximate methods introduced by earlier mathematicians were not greatly improved upon by Newton or Leibniz.

Each of these giants used the argument that higher orders of infinitesimally small increments could be neglected. The essence of their argument was that the equation

$$\Delta y = 2x\,\Delta x + (\Delta x)^2$$

would become $\qquad dy = 2x\,dx + (dx)^2$

when $\Delta x$ was infinitesimally small, and that since $dx$ itself was infinitesimally small, then a higher order – for example $(dx)^2$ – was negligible by comparison and could be ignored. Then in the transformed equation, $(dx)^2$ would be ignored before the equation was divided throughout by $dx$.

So $\qquad dy = 2x\,dx + (dx)^2$

became $\qquad dy = 2x\,dx$

and $\qquad \dfrac{dy}{dx} = 2x$

They thus arrived at the same result as is obtained by using limits, but they arrived at it in a different manner. They regarded $\dfrac{dy}{dx}$ as a true ultimate ratio despite the illogicalities involved and so left themselves open to attacks such as that made by Berkeley. This is of some interest in the development of the calculus and demonstrates the origin of the symbol $\dfrac{dy}{dx}$ which has been retained, although modern theory no longer regards it as representing a true ratio and has given it a new character in terms of limits.

Nowadays, in theory at least, we do not ignore negligible terms. We simplify the original equation as far as possible before thinking of $\Delta x$ as approaching zero. This gives us

$$\frac{\Delta y}{\Delta x} = 2x + \Delta x$$

We then say that as $\Delta x$ approaches zero, so $2x + \Delta x$ approaches $2x$.

This is the theory underlining our calculations but of course in any actual calculations we jump many of the theoretical stages. Although in explaining the theory we say that we do not ignore negligible items, it may appear in practice that we do ignore them. This however is a matter of appearance only. The pioneers of the calculus ignored $(dx)^2$ because it was negligible compared with $dx$, but this was not a sound mathematical reason.

Today in jumping from

$$\frac{\Delta y}{\Delta x} = 2x + \Delta x$$

to

$$\frac{dy}{dx} = 2x$$

we may also appear to be ignoring $\Delta x$. In so far as we may be said, in obtaining our solution, to ignore $\Delta x$ when it is made to approach zero, this is not because it is so small as to be negligible but because we are interested in the limiting value and not in the infinitesimally small increment which separates us from it.

While the increments are important as providing the main-spring of our thought, it is the derivative or the limiting value of the ratio between the increments in which the true basis of the calculus is to be found.

It will be noted that we have been able to calculate the derivative of the function $y = x^2$ entirely by analytical methods even though we referred to a geometrical figure for illustration. We shall continue to use geometrical figures where these serve to demonstrate principles, but in subsequent chapters a greater emphasis will be laid on the analytical or arithmetical approach.

*Chapter 5*

# SIMPLE DIFFERENTIATION METHODS

In the previous chapter there was developed a method of differentiating a function of the form $y = x^2$. This method is now extended to demonstrate its applicability to functions generally.

First, we take a look at the function $y = x^3$ and follow through the working exactly as before.

$$y = x^3$$

so
$$y+\Delta y = (x+\Delta x)^3$$
$$= x^3+3x^2\,\Delta x+3x(\Delta x)^2+(\Delta x)^3$$

but
$$y = x^3$$
$$\therefore\ \Delta y = 3x^2\,\Delta x+3x(\Delta x)^2+(\Delta x)^3$$

and
$$\frac{\Delta y}{\Delta x} = 3x^2+3x\,\Delta x+(\Delta x)^2$$

Now as $\Delta x$ approaches zero, each of the terms $3x\,\Delta x$ and $(\Delta x)^2$ also approaches zero, so that the limiting value:

$$\frac{dy}{dx} = 3x^2$$

Next we try the function:

$$y = x^4$$

so
$$y+\Delta y = (x+\Delta x)^4$$
$$= x^4+4x^3\,\Delta x+6x^2(\Delta x)^2+4x(\Delta x)^3+(\Delta x)^4$$

but
$$y = x^4$$
$$\therefore\ \Delta y = 4x^3\,\Delta x+6x^2(\Delta x)^2+4x(\Delta x)^3+(\Delta x)^4$$

and
$$\frac{\Delta y}{\Delta x} = 4x^3+6x^2\,\Delta x+4x(\Delta x)^2+(\Delta x)^3$$

As $\Delta x$ approaches zero, each of the terms involving $\Delta x$ (that is, each of the last three terms) also approaches zero, so that the limiting value:

$$\frac{dy}{dx} = 4x^3$$

The results achieved so far may be tabulated for the various functions:

| *Function* $y =$ | *Derivative* $\frac{dy}{dx} =$ |
|---|---|
| $x^2$ | $2x$ |
| $x^3$ | $3x^2$ |
| $x^4$ | $4x^3$ |

An emerging pattern is already discernible. Each derivative may be obtained direct from its original function by

(i) multiplying the original function by a number equal to the exponent of $x$, and then
(ii) diminishing the exponent by 1.

In the function $y = x^2$, the exponent is 2. We therefore multiply by 2, giving $2x^2$. Next we deduct 1 from the exponent, giving $2x^1$ or $2x$. This is the derivative of $x^2$. The derivatives of the other functions tabulated may be found in the same way; and, generally, the derivative of each of these functions which are all of the form $y = x^n$, may be expressed as

$$\frac{dy}{dx} = nx^{n-1}$$

This is a remarkably simple method of finding the derivative, once we know how to do it. If, for example, $y = x^{20}$ then we know at once that

$$\frac{dy}{dx} = 20x^{19}$$

whereas to carry out in full the basic calculations similar to those at the beginning of this chapter would be a most unenviable task.

The validity of the formula may be demonstrated quite easily by using the Binomial Theorem to obtain the expansion of the binomial $(x+\Delta x)^n$ where $n$ is a positive integer.

The Binomial Theorem shows that the expansion of *every* binomial of this form will consist of

(i) two terms, $x^n$ and $nx^{n-1}\,\Delta x$

and (ii) all other terms involving $(\Delta x)^2$ or some other higher power of $\Delta x$ (referred to below collectively as $A$).

An illustration of this is provided by the calculations (at the beginning of the chapter) for the functions $y = x^3$; and $y = x^4$.

In general therefore we may proceed as follows for the function $y = x^n$. Then

$$y+\Delta y = (x+\Delta x)^n$$
$$= x^n+nx^{n-1}\,\Delta x+A$$

but
$$y = x^n$$

so
$$\Delta y = nx^{n-1}\,\Delta x+A$$

In order to obtain $\dfrac{\Delta y}{\Delta x}$ we now have to divide each side by $\Delta x$. The term $nx^{n-1}\,\Delta x$ will then become $nx^{n-1}$. The terms represented by the symbol $A$ all involve $(\Delta x)^2$ or a higher power of $\Delta x$. When these are divided by $\Delta x$ therefore the quotients will all involve $\Delta x$ or a higher power of $\Delta x$. The terms so obtained may be referred to collectively as $B$. Then

$$\frac{\Delta y}{\Delta x} = nx^{n-1}+B$$

All the terms in $B$ involve $\Delta x$ and consequently, as $\Delta x$ approaches zero, all such terms (and therefore also $B$ collectively) approach zero. Therefore the limiting value:

$$\frac{dy}{dx} = nx^{n-1}$$

thus proving the validity of the formula.

The reader might like to check the instance where $y = x$. For this function it would be expected from the formula that

$$\frac{dy}{dx} = 1\,.\,x^{1-1} = x^0 = 1$$

That is, the derivative is constant at 1 for all values of $x$. This can be true only when the function is linear. The curve of $y = x$ is indeed a straight line, as may be seen in Fig. 11. The curve is in fact the hypotenuse of the triangle OBA. Consequently the gradient of the curve is $\frac{y}{x}$ and, as $y = x$, then the gradient is equal to 1. Also, in the diagram, the angle $\theta = 45°$; and tan 45°, which expresses the gradient, may be found from mathematical tables to be equal to 1.

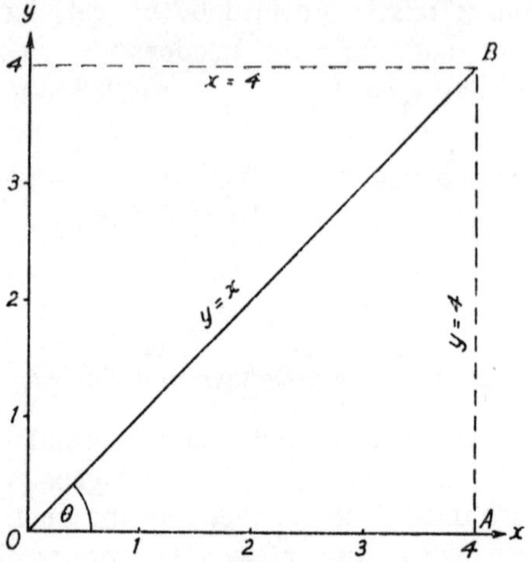


FIG. 11

The same method of differentiation holds also for negative and fractional values of $n$ (but not when $n$ is zero). Thus:

(i) $y = x^{\frac{1}{2}}$ Here $n = \frac{1}{2}$ and $\frac{dy}{dx} = nx^{n-1}$

becomes $\frac{1}{2}x^{\frac{1}{2}-1} = \frac{1}{2}x^{-\frac{1}{2}}$

(ii) $y = x^{-2}$ Here $n = -2$ and $\frac{dy}{dx} = nx^{n-1}$

becomes $-2x^{-2-1} = -2x^{-3}$

In the second instance, the derivative is a negative quantity. It has previously been said that the derivative measures the rate of growth or increase of a function. If the derivative is negative, then so is the rate of increase negative; that is, the function is in fact decreasing. An example of this is shown in Fig. 12*a* and *b*.

In curve (*a*), as $x$ is increased by $\Delta x$, so $y$ is increased by $\Delta y$. Therefore $\Delta y$ and consequently $\frac{dy}{dx}$ are both positive. In curve (*b*), as $x$ is increased by $\Delta x$, so $y$ is *decreased* by $\Delta y$. In this case therefore $\Delta y$ and $\frac{dy}{dx}$ are negative. It will also be noted that the tangent to curve ($b$) makes an obtuse angle with the $x$ axis; the tangent of this angle is therefore negative.

It was noted above that the general principle for obtaining the derivative of a function $y = x^n$ does not apply when $n = 0$. This

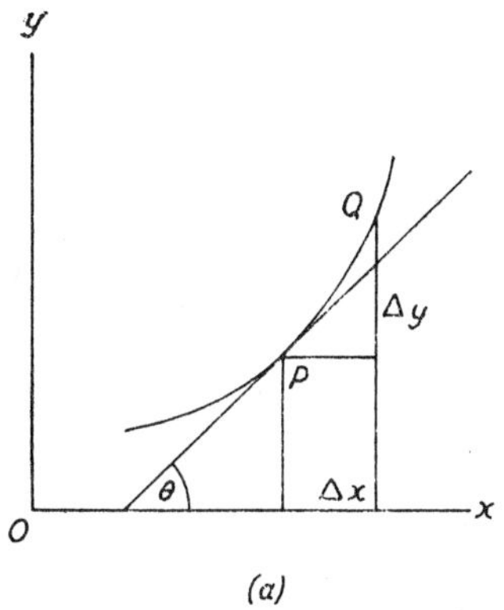


FIG. 12*a*

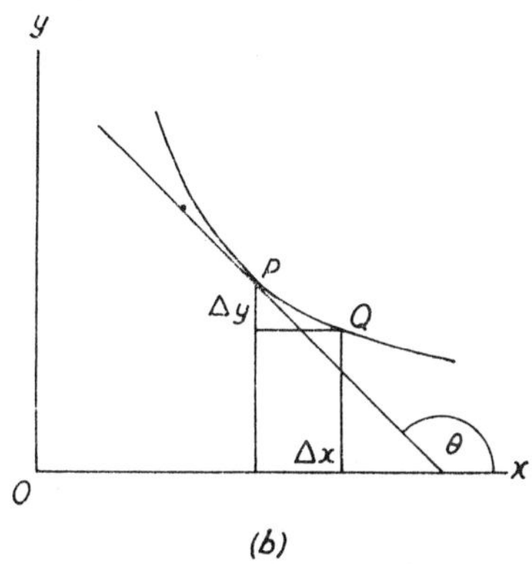


FIG. 12*b*

perhaps requires some qualification. Every number of the form $x^0$ (that is, where the exponent is zero) is equal to 1. Consequently, even though the value of $x$ may vary, the value of $x^0$ is constant at 1. There is in fact no variation to be measured and therefore no rate of change either. For the sake of consistency this might otherwise be expressed in the statement that the derivative is equal to 0. This result can actually be obtained from the general formula, thus:

$$\frac{dy}{dx} = nx^{n-1} = 0.x^{0-1} = 0.x^{-1} = 0$$

This raises an important point, for it shows that we cannot *by this method* obtain a derivative which includes the term $x^{-1}$. The exponent which remains in the derivative is always 1 less than the exponent in the function differentiated. For the exponent $-1$ to appear in the derivative the exponent in the original function would have to be 0. But we have just seen that when the exponent of the function is 0, then the derivative is itself equal to zero or perhaps does not exist at all.

From this it might seem to be impossible for $x^{-1}$ or its equivalent form $\frac{1}{x}$ to appear as part of a derivative; this is not so. The derivative $\frac{1}{x}$ may in fact be derived from a function of a different form to which reference will be made later.* One thing however is clear; it cannot be derived from a simple function of the form: $y = x^n$.

Few functions are as simple as those which have so far been considered. Many of them involve multiplying constants as well as variable quantities. In Chapter 2 it was shown that the distance fallen by a stone may be expressed as a function of time

$$y = 16x^2$$

where $y$ = distance in feet, and $x$ = time elapsed in seconds. We now proceed to find the derivative of this function and so, finally, to calculate the speed of the stone at any instant during its fall to earth.

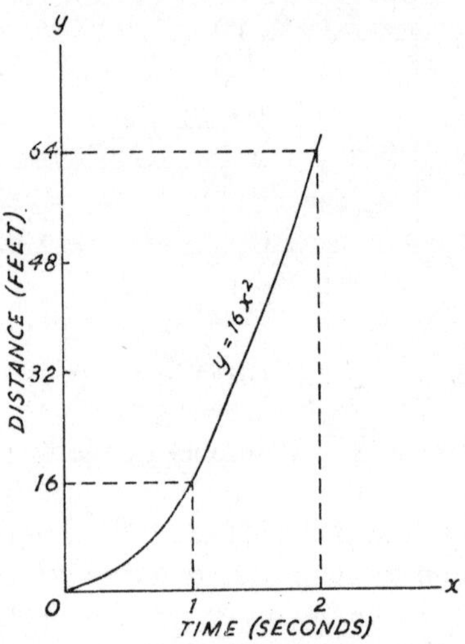


FIG. 13

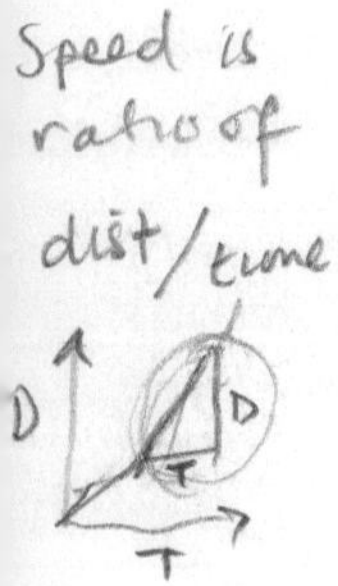


This function is shown in Fig. 13. The stone is clearly falling more and more quickly as time elapses. This is illustrated by the changing steepness of the curve. To calculate the speed of the stone we adopt the same basic procedure as we have previously used.

$$
\begin{aligned}
y &= 16x^2 \\
y+\Delta y &= 16(x+\Delta x)^2 \\
&= 16[x^2+2x\,\Delta x+(\Delta x)^2] \\
\therefore\ \Delta y &= 16[2x\,\Delta x+(\Delta x)^2] \\
&= 32x\,\Delta x+16(\Delta x)^2
\end{aligned}
$$

* See page 146.

$$\therefore \frac{\Delta y}{\Delta x} = 32x + 16\Delta x$$

As $\Delta x$ approaches zero, so also does $16\Delta x$ approach zero. Then the limiting value

$$\frac{dy}{dx} = 32x$$

The speed of the stone at any instant is therefore equal to $32x$ ft per second where $x$ is the number of seconds elapsed from the beginning of its fall. At the 5th second the stone would be travelling at $32 \times 5 = 160$ ft per sec. After only another 5 seconds (that is, at the 10th second) the speed would have doubled to 320 ft per sec. This is equal to just over 218 miles per hour – a truly remarkable speed.

For the function $y = x^2$, the derivative is $2x$. For the function $y = 16x^2$, the derivative is $32x$ or $16(2x)$. The factor 16 remains in the right-hand side of the above equations throughout the calculation and it is consequently retained unaltered as a factor of the derivative. This is because of the basic nature of the constant 16 in that by its definition it cannot vary. The variation arises only in $x^2$. The effect of the constant 16 is to magnify by sixteen times the variation originating in $x^2$. It is logical therefore that the rate of change of $y = 16x^2$ should also be sixteen times as great as the rate of change of $y = x^2$.

This kind of function may be expressed in the general form

$$y = ax^n$$

where $a$ is the multiplying constant. The differentiation procedure is then to find the derivative of $x^n$ and to multiply it by $a$ to obtain the derivative of the complete function. Thus:

$$\frac{dy}{dx} = nax^{n-1}$$

Applying this rule to the function $y = 16x^2$

$$\frac{dy}{dx} = 2.16.x^{2-1} = 32x$$

The same principle holds for 'dividing' constants, since dividing by $a$ is the same as multiplying by $\frac{1}{a}$.

Then if $$y = \frac{x^n}{a}$$

$$\frac{dy}{dx} = n\left(\frac{1}{a}\right)x^{n-1} = \frac{nx^{n-1}}{a}$$

If therefore $$y = \frac{x^2}{16} \text{ then}$$

$$\frac{dy}{dx} = \frac{2x}{16} = \frac{x}{8}$$

Different principles apply to added or deducted constants as, for example, in the function: $y = x^2+a$. Here the constant $a$ does not magnify the variation originating in $x^2$. It contributes nothing to the *growth* of the function. This may be demonstrated mathematically as follows:

Let $$y = x^2+4$$

then $$y+\Delta y = (x+\Delta x)^2+4$$
$$= x^2+2x\,\Delta x+(\Delta x)^2+4$$

but $$y = x^2+4$$

$$\therefore \frac{\Delta y}{\Delta x} = 2x+\Delta x$$

and the limiting value:

$$\frac{dy}{dx} = 2x$$

The derivative of the function: $y = x^2+4$ is thus exactly the same as the derivative of $y = x^2$. If it is remembered that the derivative measures rate of growth and also that the constant contributes nothing to the growth of a function, then the above result is quite logical. This may be demonstrated most easily by a linear function. Fig. 14 shows the curve of the function $y = x+3$. It will be seen that the effect of the constant 3 is to determine the point at which the curve cuts the $y$ axis; it has no effect on the size of the angle $\theta$. Consequently the constant can have no effect on the value of the derivative. This means that to calculate the derivative of a function involving a simple additive constant, we may ignore the constant altogether.

The same considerations apply if the constant has a negative value. That is, the derivative of the function $y = x^2 - 4$ is also equal to $2x$. It may at first seem confusing that there should be infinitely many functions each having as its derivative the value $2x$. However if the curves of all these functions could be drawn relative to the same set of co-ordinates, it would be found that

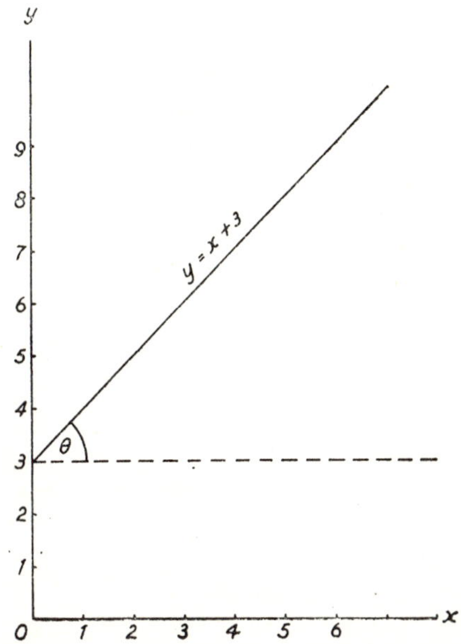


FIG. 14

the curves would always run parallel to each other. The angle made by each curve with the $x$ axis would be identical and accordingly the gradient of each curve would also be identical.

The practical implications of this are of interest. There is a police car chasing some bank robbers. The police car moves in accordance with the function $y = x+10$ (where $y$ represents distance, and $x$ represents time elapsed). The robbers' car moves in accordance with the function $y = x+15$. Can the police catch the robbers? These functions are shown in Fig. 15. Provided that the distance, $y$, is measured from the same point 0 for each car and that both cars begin to move at the same instant then the two functions show that the police car started from a point 10 units of distance away from 0, whereas the robbers started from a point 15 units of distance away from 0.

Then, if both police and robbers are travelling in exactly the

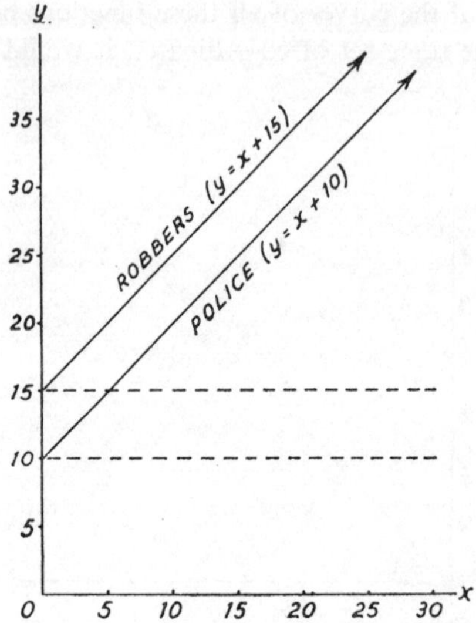


FIG. 15

same direction, the robbers would always be 5 units ahead of the police since they are travelling at the same speed. The derivative of $x+10$ is exactly the same as the derivative of $x+15$. The curves do not meet and the police do not get promotion.

## DIFFERENTIATION OF A SUM

The functions considered so far have been of the simple general form $ax^n+c$. They consist of only one term involving $x$, either on its own or in conjunction with an added positive or negative constant (for example $x$; $3x+2$; $2x-1$). Rules for the differentiation of such functions have been shown to be comparatively simple.

In practical problems however most functions will be found to contain more than one term involving $x$. If the perimeter of a rectangular field is 4 units of length and if the length of one of the sides is $x$ units, then the area ($y$) of the field, for any positive value of $x$ less than 2, is $2x-x^2$. Here $y$ is expressed as a function of $x$, but each of the terms involving $x$ contributes its own share to the growth in $y$ relative to the growth in $x$.

But these contributions are not equal and the rules for the differentiation of functions of this type must therefore provide a method for calculating the 'composite' growth in $y$. Appropriate rules for this purpose may fortunately be developed easily from those already demonstrated. Some of these developments may look a little daunting at first but the mathematics is really quite straightforward.

The derivative $\frac{dy}{dx}$ of the function $y = x^2$ is $2x$. It is now desired to find the derivative of a function expressed as a sum of terms. For this purpose we take the general function

$$y = u+v$$

or

$$y = x^3+x^2$$

where $u = x^3$ and $v = x^2$. The introduction of the symbols $u$ and $v$ helps to simplify the explanation.

The first thing to note is that as $x$ grows, so also do $u$, $v$, and $y$ grow; the growth in $y$ resulting from the individual growth in $u$ and $v$. Thus, as $x$ grows to $x+\Delta x$, so:

$u$ increases to $u+\Delta u$

$v$ increases to $v+\Delta v$

$y$ increases to $y+\Delta y$

thus $$y+\Delta y = (u+\Delta u)+(v+\Delta v)$$

but $$y = u+v$$

therefore $$\Delta y = \Delta u+\Delta v \qquad \text{(subtracting)}$$

and, dividing throughout by $\Delta x$:

$$\frac{\Delta y}{\Delta x} = \frac{\Delta u}{\Delta x}+\frac{\Delta v}{\Delta x}$$

The limiting value of the left side of the equation is equal to the limiting value of the right side, and this in turn is equal to the sum of the limiting values of $\frac{\Delta u}{\Delta x}$ and $\frac{\Delta v}{\Delta x}$.

That is $$\frac{dy}{dx} = \frac{du}{dx} + \frac{dv}{dx}$$

Here we may recognize three derivatives one of which, $\frac{dy}{dx}$, is the one we wish to identify. In the above equation this is shown to be equal to the sum of:

$\frac{du}{dx}$ (the derivative of $u$ with respect to $x$)

and $\frac{dv}{dx}$ ( ,, ,, ,, ,, ,, ,, )

This is a logical conclusion. The derivative measures the rate of growth of a function. The above equation may therefore be translated as: 'The rate of growth in $y$ relative to $x$ is equal to the sum of the respective rates of growth of $u$ and $v$ considered separately.' Thus the rate at which a whole is growing may be calculated by adding together the rates at which the parts are growing. If water is entering a tank through two separate taps at different rates of flow per minute, then the rate at which the amount of water in the tank is growing is the same as the sum of the two separate rates.

In order to calculate the derivative of the complete function we differentiate each term separately and then add together the derivatives thus obtained. Our example is

$$y = x^3 + x^2$$

The derivative of $x^3$ is $3x^2$. The derivative of $x^2$ is $2x$. The derivative of $y$ is the sum of these two derivatives. That is

$$\frac{dy}{dx} = 3x^2 + 2x$$

This principle extends to all functions of this type irrespective of the number of terms. Here are two more examples:

(i) $$y = x^3 + 16x^2 + 4x + 10$$

$$\frac{dy}{dx} = 3x^2 + 32x + 4$$

(ii)
$$y = 4x^4 + x^3 + 15$$
$$\frac{dy}{dx} = 16x^3 + 3x^2$$

In each instance the additive constant in the function has of course a zero derivative and is therefore not represented in the final derivative.

The same principle applies to negative terms. Thus if

$$y = 5x^2 - 10x$$

then
$$\frac{dy}{dx} = 10x - 10$$

The growth of $y$ due to the growth of $5x^2$ (as $x$ grows) is diminished because $10x$ also grows and the net rate of growth is the difference between the two individual rates.

## ACCELERATION

We have seen that where distance ($y$) is represented as a function of time ($x$) we may, by differentiating $y$ with respect to $x$, calculate the speed of a body. This speed may itself be varying. The falling stone is travelling at 160 ft per sec. at the end of the first 5 seconds, but this speed has increased to 320 ft per sec. at the end of the 10th second. The speed is actually varying continuously.

The rate of change of speed is more familiarly known as acceleration. It is often important to be able to calculate the instantaneous acceleration of a moving body and we must now consider how this may be derived from the original function (i.e. distance expressed as a function of time).

The speed of the falling stone at any instant is equal to the derivative of the function $y = 16x^2$. That is

$$\frac{dy}{dx} = 32x$$

If we now take this last equation on its own merits without at the moment worrying about how it was derived, we can see at once that $\frac{dy}{dx}$ varies as $x$ varies. The derivative is therefore also a function of $x$, and as such may be differentiated with respect to $x$ to derive its own rate of change. Thus if we use the symbol $s$ (=speed) to replace $\frac{dy}{dx}$ then

$$s = 32x$$

and consequently, using the general rule for differentiation:

$$\frac{ds}{dx} = 32.x^0 = 32.1 = 32$$

That is, the speed is itself changing at the rate of 32 feet per second per second.

To obtain this result we have found the derivative of the derivative of the original function. But we wish to refer our result back to the original function and we therefore say that we have calculated the *second derivative* of the original function. It will be noted that, in this instance, the second derivative is a constant (32) and does not involve $x$. The acceleration of the stone is uniform throughout its entire fall.

The process of expressing the derivative itself as a function of $x$ and then differentiating it is called *successive differentiation.* Subject to the nature of the variables involved, this process may be repeated, as may be seen with the function $y = 4x^5$.

| | | | |
|---|---|---|---|
| first derivative | $= 5.4x^{5-1}$ | $= 20x^4$ | |
| second ,, | $= 4.20x^{4-1}$ | $= 80x^3$ | |
| third ,, | $= 3.80x^{3-1}$ | $= 240x^2$ | |
| fourth ,, | $= 2.240x^{2-1}$ | $= 480x$ | |
| fifth ,, | $= 1.480x^{1-1}$ | $= 480$ | (constant) |
| sixth ,, | | $= 0$ | |

The first derivative (i.e. of the function $y$ with respect to $x$) is expressed as $\frac{dy}{dx}$. For the second derivative however it is the first derivative which is differentiated with respect to $x$. If we call the first derivative $s$, then the second derivative may be expressed as $\frac{ds}{dx}$. But $s = \frac{dy}{dx}$. Therefore the second derivative may be expressed as $\frac{d\left(\frac{dy}{dx}\right)}{dx}$. This form of expression is too clumsy for general use and, similarly, the expressions for higher or successive derivatives would be even more clumsy. The expression for the second derivative is accordingly simplified to $\frac{d^2y}{dx^2}$. The third derivative is shown as $\frac{d^3y}{dx^3}$, the numerals used in the derivatives providing the clue as to which derivative is intended.

A word of caution is necessary here. The expression $\frac{d^2y}{dx^2}$ does not mean '*d* squared times *y*, divided by *d* times *x* squared'. The expression must be treated as being one composite symbol – the symbol representing the second derivative and nothing else.

Even this notation is regarded as clumsy and it is useful to reintroduce the more generalized form of notation. In this, the symbol $f$ is used to mean 'function of' (as in $y = f(x)$ meaning '$y$ is a function of $x$'). The symbol does not identify the actual relationship between $y$ and $x$ but merely indicates its existence. The first derivative is then expressed as $f'(x)$, merely by the addition of one dash following the $f$ symbol. For successive higher orders of derivative, a further dash is added at each stage. Derivatives are thus represented as follows:

$$y = f(x)$$

| | |
|---|---|
| | $y = f(x)$ |
| first derivative | $= f'(x)$ |
| second ,, | $= f''(x)$ |
| third ,, | $= f'''(x)$ |

and so on.

In so far as we have been dealing with distance expressed as a function of time, it has been comparatively easy to give everyday meanings to the first and second derivatives – speed and acceleration respectively. Furthermore, if instead of referring to speed we refer to rate of change, and if instead of referring to acceleration we refer to the rate at which the rate of change is itself changing, then we may extend our understanding of the first and second derivatives to all functions. Reference has been made however to third, fourth, and even higher successive derivatives. How are these to be interpreted in the physical sense?

There is no simple answer to this, for the very good reason that the applications of the higher order derivatives are themselves not simple; to understand the applications it is first necessary to have an understanding of the branch of science in which they are being applied. We do not therefore propose to attempt to explain such applications here. It will be time enough to learn of these when the need arises. The reader may rest assured however that higher derivatives are essential to the understanding of many of the more advanced branches of science; and as just one example of physical application, they are very important in the study of the forces of shear and stress in structural engineering.

It is always useful to have a physical or concrete application or interpretation whereby we may seek to illustrate a mathematical concept or theory. But here again it must be remembered that although mathematical thought may be rooted in experience, nevertheless much of its development is very often abstract. If the variable $y$ is a function of $x$, then the derivative is also a function of $x$. Mathematicians, having once made this discovery, then realized that the process could be repeated. Whether the higher order derivatives have any physical application at all is of no prime importance to the research mathematician. He is concerned more with the mathematical system itself.

There are many similar instances in mathematics. It is accepted that the number $x^2$ may be represented by a square each side of which is $x$ units in length. But we cannot in the same way draw a figure in two dimensions to represent $x^{20}$ since this has twenty 'dimensions'. Nevertheless the number $x^{20}$ can have a very definite meaning and useful application. Higher powers of numbers occur frequently in applied mathematics; so also do higher order derivatives.

Bishop Berkeley in his famous criticism sought to counter the assertions of an 'infidel mathematician' (to whom the tract is addressed) that many of the doctrines of Christianity were 'inconceivable'. He sought to show that the mathematical ideas of the calculus which, he implied, were devoid of physical application, must also necessarily be inconceivable. Referring to Newton's system of fluxions* Berkeley wrote:

But the velocities of the velocities . . . [i.e. rates of change of the

* See page 69.

> derivatives]—the second, third, fourth and fifth velocities . . . exceed, if I mistake not, all human understanding. . . . Certainly, in any sense, a second or third fluxion [i.e. derivative] seems an obscure Mystery. . . . And if a second fluxion be inconceivable, what are we to think of third, fourth, fifth fluxions and so on without end? . . . But he who can digest a second or third fluxion, a second or third difference, need not, methinks, be squeamish about any point in divinity. . . .

In this tract Berkeley was of course scornfully telling mathematicians who lived in a mysterious world of their own not to criticize the mysteries of other people's worlds. As has previously been mentioned, this tract did much to call attention to the need of a more rigorous formulation of the calculus. Yet nowadays we are at least one up on Berkeley; we can easily achieve an understanding of acceleration as an example of a second derivative.

*Chapter 6*

# PAUSE FOR A PARADOX

It is now desirable to take a short pause to consider certain questions of logic which the last chapter may have posed.

Comment has already been made on the difficulties which arise in trying to reconcile intuitive ideas with abstract mathematical concepts. Intuition is more often than not a most unreliable guide to mathematics but some attempt at reconciliation between the two points of view is always worthwhile. The difficulty of course is that if an idea is mathematically abstract it can be satisfactorily described only in the language of mathematics. Any down-to-earth explanation which makes an appeal to intuition must be accompanied by the warning that it is offered as an aid to understanding and not necessarily as defining the mathematical idea.

In the last chapter we claim to have calculated the instantaneous speed, or speed at any instant, of a falling stone. Although it has been demonstrated that the value obtained for the speed is numerically exact, we may not as yet be clear as to precisely what it is that has been measured. What *is* instantaneous speed?

When we come to consider the interpretation of this term we encounter a number of awkward questions which appear to reveal some very inconvenient contradictions. These questions may be summarized as follows.

The values of $x$ and $y$ at any point on a curve are fixed relative to each other. How then can one be said to be varying with respect to the other at that point? How can there be a rate of change or a speed at that point? It is quite obvious that the stone has a speed, but how can it?

The term 'instantaneous speed' is perhaps a misleading one in so far as it might seem to suggest that a speed may exist in respect of a non-existent interval of time, or of an interval of time which has no dimensions. Intuition regards an instant rather as something which passes in a flash but which is nevertheless finite. To calculate the speed or change, however small, *during* an interval obviously requires that there shall be an interval, however minute it may be. In mathematics instantaneous speed is regarded as the limiting value of a variable *at* a point.

This raises another question. If speed is related to a point which geometrically has no dimensions then how can the instant itself have any dimensions? And, if it has none, how can it be used as a basis for a rate of change calculated as a ratio between dimensions?

These questions appear to present a pretty pickle but appropriately there are one or two red herrings among them. Some of the difficulties arise because of the way in which the questions are worded.

An instant is a mathematical abstraction just as is a point and, as with the latter, it may help us to identify and deal with a mathematical problem which is hidden within a physical problem. If we are to understand the former in terms of the latter then even if it is correct to regard the values of $x$ and $y$ as fixed at any point on the curve, we must take into account the nature of the variables involved. If there are physical quantities involved then the change of value of the variable may be interpreted as the change in some measurable physical attribute. If we talk of speed, then there must be something to which the speed belongs.

The distance travelled by the stone, the time elapsed since it started to fall, its speed and its acceleration are all variables measuring some attributes of the stone. They are all related to each other, but each of them in turn may be regarded as an independent variable. We express distance as a function of time, but we could just as well express time as a function of distance. We usually regard speed as a function of time and distance; and we tend to think of it as depending upon these. This however is not necessarily true. The stone cannot attain a stated speed unless it covers a certain distance in a certain time. But equally it cannot cover that distance in that time unless it attains the stated speed.

A reasonable extension of this idea allows us to regard speed as existing in its own right as an independent variable. This is not far removed from the intuitive idea of speed. A motorist instinctively talks about the speed of his car without consciously thinking of it as a ratio between distance and time.

If we say that the values of distance and time ($y$ and $x$) are fixed for any point on the curve, then we are equally justified in saying that the speed ($s$) and the acceleration ($a$) also have fixed values at that point. These 'fixed' values do not refer to the performance of the stone *within* any interval. Instead they indicate the values of the variables *at* that point, and nothing else.

Speed and acceleration may be expressed as functions of time and we may draw their curves relative to the same co-ordinate axes as we use for the original curve of $y = 16x^2$. These are shown in Fig. 16*a*. If we think of the whole diagram as if it were

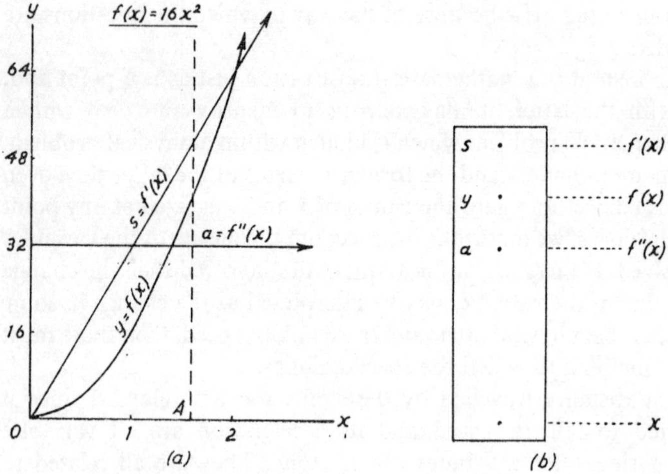


FIG. 16*a*. CURVE OF A FUNCTION AND ITS FIRST AND SECOND DERIVATIVES

FIG. 16*b*. CROSS-SECTION DRAWN AT POINT A

a thin slab of concrete, then the curves may be thought of as being represented by thin wires set in the concrete. If we then cut the slab at A and examine its cross-section, the latter would look something like Fig. 16*b* (rather like currants in a poor fruit cake). This cut merely reveals the position of the wire cross-sections (or the position of the currants in the cake) relative to each other. Selecting any instant at which to inspect the values of any of the variables is rather like making the cut and examining the cross-sections represented by points.

In Fig. 16*a* the values are all measured from the origin of the co-ordinate axes and they thus represent the values attained by each of the variables at the points in their respective curves relative to the value of $x$ at point A. They do not measure anything which has occurred either at or during the point but instead are indicators of what has been attained up to that point. Whether the point or instant has dimensions is therefore irrelevant to the calculation of speed. The instantaneous speed of the stone may be regarded, not as the ratio between distance and time within an instant but, instead, as the speed which has been attained by the stone at that instant.

It will probably help to reconcile the mathematical and the intuitive ideas if we understand that instantaneous speed is the limiting value of the average speed over an interval as that interval is continuously decreased. This in fact is how it is calculated since the ratio $\frac{\Delta y}{\Delta x}$ (the limiting value of which measures the speed at any instant) represents the average speed over the time interval $\Delta x$.

This whole matter is of course closely bound up with the subject of continuous variation. To be able to pose the questions at all we have had to accept the theory that a continuous variable can have a 'static' value and that its curve is an infinite aggregate of points.

As was shown in Chapter 2, the basic idea used here of a continuous variable is essentially a mathematical one. It is defined entirely in numbers. Most people have some idea of continuity based on the idea of physical motion, or at any rate on ideas associated with the elapsing of time, but they would not be able to define it except in terms of itself. Newton's own concept of the continuity of a variable was of a *flow* and he drew his terms from the same idea. The variable was called a *fluent*, and the quantity which we now call the derivative was said to measure the rate of flow and was called a *fluxion*. But the ordinary idea of continuity is too vague to act as a basis for mathematics. The mathematical idea is a precise one even if it cannot be fully grasped intuitively.

We say that a continuous variable assumes all possible values between $a$ and $b$. The trouble is that we cannot see it doing so. We have to accept that it does act in this way largely because we cannot see how it can act in any other way. If we travel in a train through a tunnel, we know when we emerge at the other end that we have passed over every possible subdivision of the railway track. During the journey we would not have been aware of these divisions as physical entities. We could not have seen them any more than we could see a variable assuming all its intermediate values. But we do know that if there were some 'possible' divisions over which we had not travelled, then we could not have completed the journey.

The apparent paradox involved in this discussion is one which would also seem to deny the possibility of motion, let alone the possibility of the speed accompanying that motion. It is thus basically the same as that which was the subject of another of Zeno's paradoxes – the arrow in flight.

The argument of the parodox is stated thus: the arrow in flight is *moving* during every instant, yet at every instant it must be occupying a *specific* position in space. These propositions contradict each other. If an object occupies a specific position (that is, it is *somewhere*) then it is temporarily stationary; it is not changing its position and it is not therefore moving. So much for the paradox; where is the fallacy?

Motion is change of position and this is measured as distance travelled. Distance or length is a continuous variable, as also is time, and the mathematical view of continuity enables us to regard each of them as composed of an infinity of points or instants. We may match up pairs of values of the two variables and regard these as fixed or static with regard to each other *only* if we first accept that the continuous variables may be regarded as a totality of points. We cannot then use the existence of static values to deny the continuity of the variables.

Modern theory generally accepts that a moving body is always at rest in the sense applied above. This might suggest that mathematics is trying to have the best of both worlds, but of course this is not so. Motion is regarded merely as the representation of a rapid succession of an infinite number of changes of position. In fact the idea of the existence of fixed values at a point is possible because of and not in spite of the fact that there is an infinity of such pairings.

If all this theory is too much to take, then perhaps the best thing is to go out to the cinema! There you can take it easy watching your favourite stars move across the screen – but, wait, you are not seeing anything of the sort! What you are seeing is the projection of a succession of still photographs. If you were to study these on the reel you would find that each is slightly different from the rest in that the actors have assumed slightly different positions. The speed at which these photographs were originally taken enables such very small differences to be distinguished and filmed. The speed at which the completed film is then run in the projection box helps to merge the separate photographs together so that they appear on the screen as a continuous film built up of static images.

We in turn break a continuous variable up into points or instants or fixed values – call them what we will – make our calculations, and then refer these back to the original continuous variable. The continuity of the variable itself is not disturbed by the devices we use in the intermediate stages.

# Chapter 7

# MAXIMUM AND MINIMUM VALUES

In dealing with the derivative of a function we have represented this primarily as a measurer of speed simply because speed is a concept which is relatively easy to understand. At the same time it has been stressed that speed is only a particular form of a rate of change and we now consider the more general aspect of the derivative in problems which do not involve speed or physical motion.

One of the major applications of differentiation is in the determining of maximum or minimum values; that is, in finding the best values for specific purposes. It should be said at once that in this context the words *maximum* and *minimum* have a restricted meaning, the exact nature of which will become evident as we proceed.

This aspect of differentiation is an extremely important one in many scientific and engineering applications. The sort of problem which it helps to solve is the determining of what conditions (that is, what value of an independent variable) will provide a maximum efficiency in the operation of a machine or will make the operating cost a minimum.

In a simpler but by no means trivial example we may solve the problem of a manufacturer who wishes to pack detergent powder in cardboard cartons. He has a large stock of cardboard pieces which he wishes to use. Each piece is 12 inches by 12 inches. The cartons are to be made by cutting smaller squares out of the corners of the cardboard pieces (as in Fig. 17), so that the sides of the cartons would be formed by folding the remaining pieces along the dotted lines. What is the maximum volume of powder which may be packed in each carton and what is the size of the squares which must be cut out of the pieces of cardboard to achieve this volume?

This is a very simple problem to solve once we can handle the differential calculus. The first step is to identify the function involved. Let the volume of the carton be $y$ cubic inches, and let the length of each side of the corner squares cut out be $x$ inches. When the cardboard is folded along the dotted lines, the resultant carton will then have a height of $x$ inches. The base of the carton will be a square of side $(12-2x)$ inches, so that the area of the base will be $(12-2x)^2$ square inches. Consequently the volume of the box will be $x$ multiplied by $(12-2x)^2$. That is,

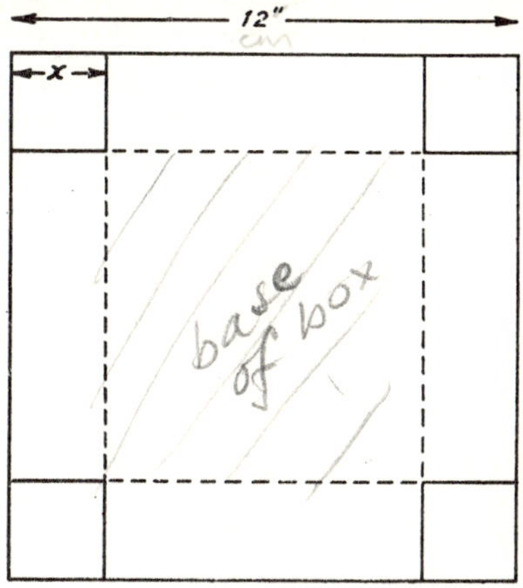


FIG. 17

$$y = 4x^3-48x^2+144x$$

For what value of $x$ will this function be a maximum value?

Before solving this by differentiation it is desirable to discover the behaviour of $y$ as different numerical values are substituted for $x$. These may be tabulated as

| | | | | | | | |
|---|---|---|---|---|---|---|---|
| $x =$ | 0 | 1 | 2 | 3 | 4 | 5 | 6 |
| $y =$ | 0 | 100 | 128 | 108 | 64 | 20 | 0 |

and these values are plotted on the curve in Fig. 18.

From this it appears that the highest value of $y$ occurs when the value of $x$ is about 2. We cannot say that it occurs when $x = 2$ exactly. It is evident that $y$ continues to increase up to a point relative to a value of $x$ between 1 and 3. But we have plotted only integral values of $x$. If we now take $x = 2{\cdot}25$, then $y = 126{\cdot}5625$. Similarly if $x = 1{\cdot}75$, $y$ is again less than 128 and it appears probable that $x = 2$ really does give a maximum value for $y$. This is still not absolutely certain however merely by reference to the curve. The calculus will provide this certainty.

Hitherto we have talked mainly of functions growing or increasing, and the curves of these functions continue on an ever upward course across the squared paper. Furthermore, the derivatives of these functions are always positive. In Fig. 18 however is a curve which behaves differently. At first the value of $y$ increases as $x$ increases, but, having reached some maximum value, it then decreases. The curve rises and then falls. It will at once be obvious that the function must stop increasing before it can decrease and that the highest value of the function occurs at the point where the growth stops immediately prior to its fall. This highest value is called a *maximum value* of the function and the point where this value occurs on the curve is called a *maximum point*.

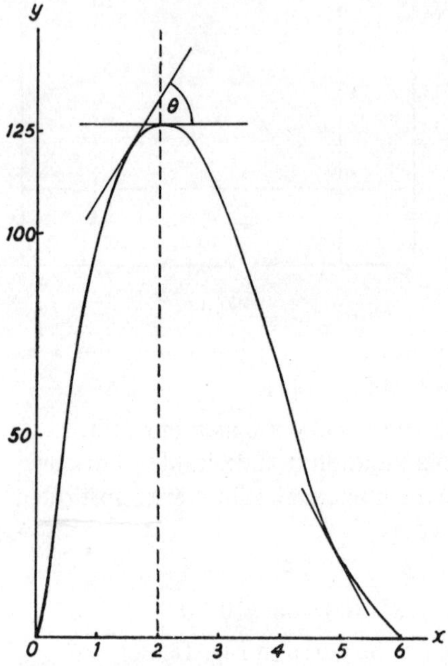


FIG. 18

The derivative of a function is a measurer of the rate of increase of the function if it is positive in value. If the derivative is negative in value however it instead measures the rate of decrease. Geometrically the derivative may be represented as tan $\theta$ when $\theta$ is the angle made between the axis and the tangent to the curve (as shown in Fig. 18 in respect of the tangent drawn at the point $x = 1{\cdot}5$). In this instance the tangent may be described as rising from left to right. At the point $x = 5$ however, where the function is decreasing, the tangent may be described as falling from left to right. The angle $\theta$ in this instance is obtuse and tan $\theta$ is therefore negative.

The sign of the derivative at any point of the curve thus indicates whether the value of the function is increasing or decreasing at that point. If the sign is positive, then the function is increasing. If the sign is negative, then the function is decreasing. But at a maximum point the function is temporarily neither increasing nor decreasing, just as the mountaineer at the peak of a mountain is neither climbing nor descending. At this point therefore the rate of change is temporarily zero and the derivative at this point is also zero, which is represented in Fig. 18 by the gradient of the tangent at that point. This tangent is in fact parallel to the $x$ axis and the angle $\theta = 0$.

The derivative at a maximum point is equal to zero. That is

$$\frac{dy}{dx} = 0$$

If therefore we find the derivative of the function and equate this derivative to zero, the equation thus set up will enable us to calculate the value of $x$ at the point where $\frac{dy}{dx} = 0$. The function which expresses the volume of a cardboard carton is

$$y = 4x^3 - 48x^2 + 144x$$

Proceeding by basic principles, then

$$\frac{dy}{dx} = 12x^2 - 96x + 144$$

But this is equal to zero at a maximum point

$$12x^2 - 96x + 144 = 0$$

or

$$x^2 - 8x + 12 = 0 \qquad \text{(dividing by 12)}$$

$$\therefore (x-6)(x-2) = 0$$

$$\therefore x = 2 \text{ or } 6 \quad \text{(inches)}$$

The value of $x$ cannot be 6 inches because of the restrictions of the problem. If we try to cut $6'' \times 6''$ squares out of the corners of a $12'' \times 12''$ piece of cardboard, we will finish up with just four squares of cardboard and no box. Consequently the value of $x$ which makes $4x^3 - 48x^2 + 144x$ a maximum is $x = 2$.

This is the result which we previously suspected to be correct. The difference now is that we know that the result is exact. There is no need to substitute, say, $x = 2{\cdot}01$ to see if this gives a greater volume; we know definitely that it cannot. By the original method of tabulation it was first necessary to calculate separate values of $y$ for a number of different values of $x$ and then to select the apparent highest value by inspection. Using the calculus we can find the exact value with a great economy of working.

Some comment here on the equating of $\frac{dy}{dx}$ to zero may be useful, since the logic of doing so is sometimes doubted.

We have said that the derivative may be regarded as an instantaneous rate of change or speed. But if a car is moving, for example, how is it possible for it to have a zero speed at any point? The short answer to this is that it cannot! Speed is the rate of change of absolute mileage travelled irrespective of direction. If this mileage is represented as a function of time and its curve is drawn, then the doubt expressed above will not arise for the curve cannot have a maximum point. While a car is moving its absolute mileage is always increasing and the curve will continue to rise.

If a curve representing distance as a function of time does have a maximum point, this is only possible if the distance is measured in a specific direction from a starting point. When the curve turns back towards the $x$ axis, the car is travelling back towards its starting point. The car has gone into reverse. It has to stop, even though only instantaneously, before it can retrace its path. In that instant its speed is zero; the distance is neither increasing nor decreasing. Speed in a specific direction is called velocity and it is always necessary to distinguish between this and mere speed.

In developing the geometrical interpretation of $\frac{dy}{dx}$, it was emphasized that this is the limiting value of $\frac{\Delta y}{\Delta x}$ as $\Delta x$ approaches zero; and that $\frac{dy}{dx}$ itself is not zero but has some value, this value being different at different points of a curve according to the ascending or descending direction of the curve at that point. It does not follow however that this value cannot be zero at a particular point. If a continuous function passes from a positive to a negative value as $x$ varies, then it must necessarily pass through some point at which its value is zero since it must pass through every intermediate value.

Just as every function which first increases and then decreases must have a maximum value, so every function which first decreases and then increases must have a *minimum value*. The point on the curve at which this minimum value occurs is called a minimum point. Minimum and maximum points are collectively called *turning points*.

Fig. 19 illustrates the curve of $y = x^2 - 4x + 8$. At point A the curve stops descending and begins to ascend. This is the reverse of the behaviour of a curve at a maximum point but in each case, whether at a minimum or maximum point, the function is momentarily neither increasing nor decreasing and the tangent to the curve at that point is parallel to the $x$ axis.

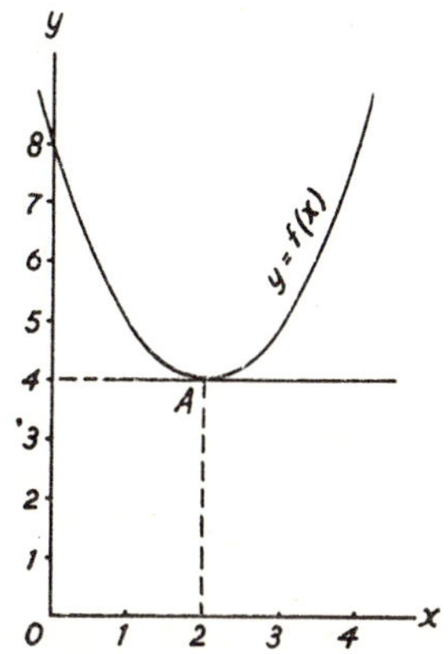


FIG. 19

This immediately raises a question. If $\frac{dy}{dx} = 0$ at a maximum and also at a minimum point then, when we equate the derivative to zero, how do we know if the resultant value is a minimum or a maximum? There is more than one way in which this may be answered. If the curve has been drawn, the facts will be obvious by inspection. Again, the facts may be clear by inspection even where no curve has been drawn; some results will be evident from the nature of the function. Where there is no self-evidence however some general rule is required. The appropriate rule is very simple. It is only necessary to check whether the function is increasing *before* or *after* the turning point or, alternatively, whether it is *increasing* or *decreasing* before the point. If it is increasing before the turning point, then this point is a maximum; if it is decreasing before the turning point, then the point is a minimum.

This rule is sufficient if it is known that the point is either a maximum or a minimum point. If this is not known, it is necessary to check both before and after the point to observe the behaviour of the function.

If the function is increasing, then $\frac{dy}{dx}$ must be positive. If the function is decreasing, then $\frac{dy}{dx}$ must be negative. This is represented in Fig. 20 by drawing tangents at A (before the maximum point), at B (the maximum point), and at C (after the maximum point). The value of $\frac{dy}{dx}$ is positive for all points between A and B, but the *rate* of increase of the function is falling off – the curve becomes less steep. That is, $\frac{dy}{dx}$ is decreasing at a point just before the turning point. At the turning point, $\frac{dy}{dx}$ is zero. After the turning point, $\frac{dy}{dx}$ is negative. Thus the rate of increase falls off before the turning point and it continues to decrease after the turning point. That is, $\frac{dy}{dx}$ is decreasing at every point between A and C. Most important of all is the fact that $\frac{dy}{dx}$ is decreasing at the maximum point.

In Chapter 5 it was shown that the second derivative $\frac{d^2y}{dx^2}$ measures the rate of change of the first derivative $\frac{dy}{dx}$. The sign of $\frac{d^2y}{dx^2}$ will therefore indicate how $\frac{dy}{dx}$ is itself changing. At a maximum point $\frac{dy}{dx}$ is decreasing; consequently its rate of change, that is $\frac{d^2y}{dx^2}$, must be negative. It follows that once we have calculated a turning value, all that is then necessary is to calculate the second derivative. If this is negative then we know that the turning value is a maximum value.

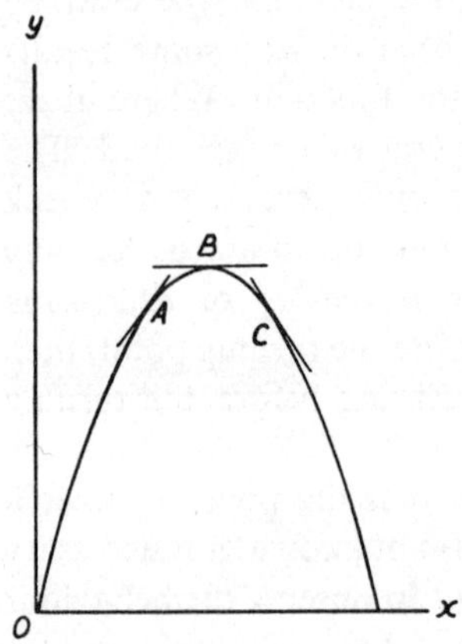


FIG. 20. TANGENTS DRAWN AT AND NEAR TO A MAXIMUM POINT

The converse is true of a minimum value. If a function is decreasing before and increasing after the turning point, then

$\frac{dy}{dx}$ is negative before and positive after the point. That is, $\frac{dy}{dx}$ is increasing at the turning point and therefore $\frac{d^2y}{dx^2}$ must be positive.

There is accordingly a very simple rule. At any turning point,

$\frac{d^2y}{dx^2}$ is NEGATIVE for a MAXIMUM point
POSITIVE for a MINIMUM point

If we now revert to the problem of a maximum value of carton volume, we have

$$y = 4x^3 - 48x^2 + 144x$$

$$\frac{dy}{dx} = 12x^2 - 96x + 144$$

$$\frac{d^2y}{dx^2} = 24x - 96$$

At the turning point, $x = 2$. Therefore the value of $\frac{d^2y}{dx^2}$ at this point is $(24 \times 2) - 96 = -48$. Since this is negative, the turning point is a maximum point.

Before proceeding further it is perhaps desirable to refer to a question which calculus students sometimes ask. The second derivative is derived from the first derivative. If we equate the first derivative to zero in order to calculate the turning value, how then can the second derivative have any value other than zero? How then can the question arise as to whether the second derivative is negative or positive?

This question confuses the general nature of a derivative with its particular value at a specific point. The first derivative describes the behaviour of the original function. Thus the first derivative of a function which expresses distance in terms of time represents speed. But the second derivative describes the behaviour of the first derivative; it shows the rate at which the value of the first derivative is itself varying.

The first derivative is equated to zero only in respect of a specific point on the curve of the original function. When $\frac{dy}{dx}$ is zero, the curve of the function is passing through a turning point. But $\frac{dy}{dx}$ is itself a function of $x$ and its values may be represented by a curve in exactly the same way as values of other functions. The second derivative $\frac{d^2y}{dx^2}$ measures the gradient of this second curve. Thus the value of $\frac{d^2y}{dx^2} = 0$ when the curve of $\frac{dy}{dx}$ is passing through a turning point, and not necessarily when the curve of $y$ is passing through a turning point. The equating of $\frac{dy}{dx}$ to zero does not also automatically equate $\frac{d^2y}{dx^2}$ to zero.

Is this not opposed to the rule for differentiation? If, for example, $\frac{dy}{dx} = 12$ (i.e. constant) then $\frac{d^2y}{dx^2} = 0$ since $\frac{dy}{dx}$ has no rate of change. If in fact $\frac{dy}{dx} = 0$ (i.e. less than 12), how can the numerical value of $\frac{d^2y}{dx^2}$ be greater than zero?

This is where the confusion occurs. Consider the function $y = 12x$. Then $\frac{dy}{dx} = 12$ for all values of $x$; it is a constant value and accordingly $\frac{d^2y}{dx^2} = 0$. There is no change in the value $\frac{dy}{dx}$ whatever the value of $x$.

Now consider the function $y = 12x-x^2$. Then $\frac{dy}{dx} = 12-2x$. In this instance $\frac{dy}{dx}$ is not a constant value. It varies as $x$ varies. When $\frac{dy}{dx}$ is equated to zero, this is effected only in respect of a specific value of $x$ which we wish to identify. In doing this we are not therefore implying that $\frac{dy}{dx} = 0$ for every value of $x$. In fact if $\frac{dy}{dx} = 0$ for every value of $x$ then there can be no rate of change in the function $y$. That is, $y$ must be a constant value and its curve will be a straight line.

It is instructive to draw the curve of a function and then, on the same scale and between the same axes, to superimpose the curve of the derivative. Fig. 21 shows the curve of $y = x^2 - 4x + 8$; and also the curve of its derivative: $\frac{dy}{dx} = 2x - 4$. It will be noted that the curve of the derivative cuts the $x$ axis at the value of $x$ which gives a turning point on the curve of the original function $y$. That is, it cuts the $x$ axis at $x = 2$. This agrees with the calculation

$$y = x^2 - 4x + 8$$

$$\therefore \frac{dy}{dx} = 2x - 4 = 0 \qquad \text{(for a turning value)}$$

$$\therefore \quad x = 2 \qquad (\text{,,} \quad \text{,,} \quad \text{,,})$$

It has been noted that a function increases before a maximum point and decreases afterwards; conversely, a function decreases before a minimum point and increases afterwards. These forms of behaviour of a function are all that are implied in the words *maximum* and *minimum*. The use of the word *maximum*, for example, does not mean that any value which we calculate as a maximum value of a function is necessarily the greatest possible value of that function. All that it implies is that the function is increasing up to that value and that it decreases afterwards. The apparent paradox here arises only out of the restriction imposed upon the meaning of the word.

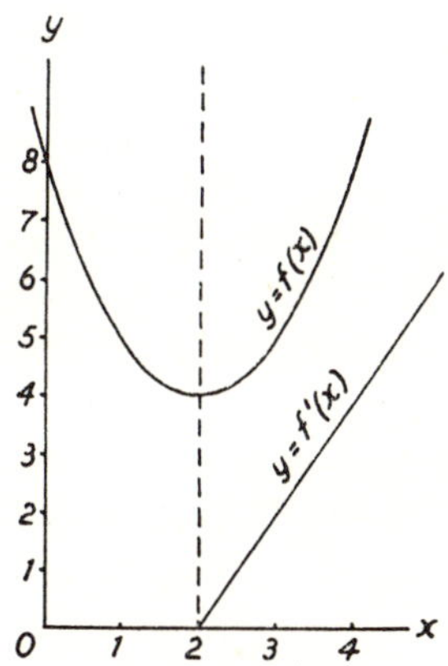


FIG. 21. CURVE OF A FUNCTION AND OF ITS DERIVATIVE

The position will become clearer however when it is realized that a function of a particular form may have more than one turning value, and yet have no finite greatest or least value. A curve of the kind is shown in Fig. 22 where

$$y = x^3 - 6x^2 + 11x - 6$$

This function may be factorized to

$$y = (x-1)(x-2)(x-3)$$

and it is clear from this that $y = 0$ when $x = 1$, 2, or 3. Consequently the curve will cut the $x$ axis at each of these three points.

From the diagram it is also apparent that there must be a

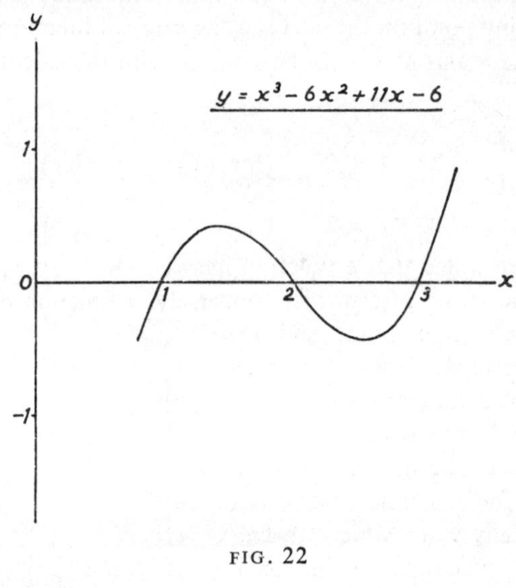


FIG. 22

maximum point between $x = 1$ and $x = 2$; and there must be a minimum point between $x = 2$ and $x = 3$.

The turning points are identified as follows:

$$y = x^3 - 6x^2 + 11x - 6$$

$$\therefore \quad \frac{dy}{dx} = 3x^2 - 12x + 11 = 0 \qquad \text{(at turning points)}$$

whence $x = 1{\cdot}42$ and $2{\cdot}58$ (approx.)

Further $\dfrac{d^2y}{dx^2} = 6x-12$

$$= -3{\cdot}48 \qquad \text{when } x = 1{\cdot}42$$
$$= +3{\cdot}48 \qquad \text{when } x = 2{\cdot}58$$

The value of $y$ relative to $x = 1{\cdot}42$ is a maximum value since the second derivative is negative for this value of $x$. When $x = 2{\cdot}58$, the second derivative is positive and the value of $y$ relative to this value of $x$ is a minimum value.

If the part of the curve between $x = 1$ and $x = 3$ is now ignored, then it may be proved that throughout the rest of the curve the function increases or decreases indefinitely. Although

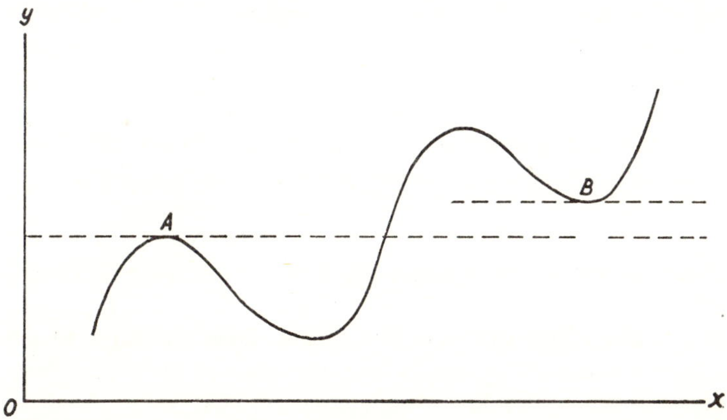


FIG. 23

the value of $y$ is a maximum when $x = 1{\cdot}42$, it is not the greatest possible value of $y$.

A maximum value on the curve of a function is therefore defined as a value such that the values immediately preceding it and immediately following it are less than the maximum. Conversely a minimum value is a value such that the values immediately preceding it and following it are greater than the minimum. The terms maximum and minimum are restricted to those parts of the curve where the circumstances they describe are located. That is, they have a local significance and not a general one. In the same way the oldest inhabitant of London may be 90 years old, whereas the oldest inhabitant in the whole of Britain may be 105 years old.

Furthermore a maximum point in one part of a curve does not necessarily represent a value of $y$ greater than the value of $y$ at a minimum point in another part of the curve. An example of such a curve is given in Fig. 23. The value of $y$ at point A is a maximum. The value of $y$ at point B is a minimum. Nevertheless the value of $y$ at B is greater than the value at A.

The restriction in the meaning of maximum and minimum values may appear to introduce difficulties of recognition. If it is required to calculate the greatest value of a function for a specific purpose and a maximum value is duly found, how is it possible to tell that this is not just one of a number of maximum values and thus possibly not the value which is required? The clue here is in the words *for a specific purpose*. The conditions of a problem will normally erect limits to the values of the function and this will enable us to concentrate our attention to a specific part of the curve.

Further examples of the application of these principles now follow (and an additional example is given in Chapter 10).

### MAXIMUM AREA OF A RECTANGLE

If the perimeter of a rectangle is 6 inches, what is the maximum area which can be enclosed within this boundary? Let the length of one side of the rectangle be $x$ inches. Then the length of the adjacent side is $3-x$ inches and the area of the rectangle is therefore $x(3-x)$ or $3x-x^2$ square inches. That is

$$y = 3x-x^2$$

$$\therefore \frac{dy}{dx} = 3-2x = 0 \qquad \text{(for a maximum)}$$

$$x = 1\tfrac{1}{2} \qquad (\text{,, ,, ,,})$$

and

$$3-x = 1\tfrac{1}{2}$$

Accordingly the maximum area is achieved when the adjacent sides are both $1\frac{1}{2}$ inches; that is, the maximum area occurs when the rectangle is in fact a square. This applies to any rectangle. If the perimeter be $k$ inches, then $y = x\left(\frac{k}{2}-x\right)$ or $\frac{kx}{2}-x^2$ and

$$\frac{dy}{dx} = \frac{k}{2}-2x$$

Equating this to zero for a maximum gives

$$2x = \frac{k}{2} \qquad \text{whence } x = \frac{k}{4}$$

Thus no matter what value we assign to $k$ as the perimeter length, each side is a quarter of the perimeter. Two adjacent sides of a rectangle together sum to one half of the perimeter. Therefore the adjacent side is also a quarter of the perimeter.

The fact that the value $x = \frac{k}{4}$ gives a maximum value of $y$ is checked by calculating the second derivative

$$\frac{dy}{dx} = \frac{k}{2} - 2x$$

$$\therefore \frac{d^2y}{dx^2} = -2$$

This is negative and the value of $y$ is therefore a maximum.

The calculation of the maximum area of a rectangle played a fairly prominent part in mathematical thought leading to the calculus.

Pierre de Fermat, a French mathematician of remarkable ability, had evolved a method which in essence is almost the same as that given by the differential calculus.

If one side of the rectangle is $x$ then the area, as we have already seen, is $x(3-x)$. This is assumed to state a specific value of $x$ which would make the area a maximum. If this value is slightly increased by an amount $E$, then the value $(3-x)$ must become $(3-x-E)$. The new area will then be $(x+E)(3-x-E)$. If $E$ is very small then the original area and the new area will be approximately equal. That is

$$x(3-x) = (x+E)(3-x-E) \qquad \text{(i)}$$

is approximately true. Equation (i) can be simplified to

$$3x-x^2 = 3x-x^2-Ex+3E-Ex-E^2$$

whence $\qquad 0 = E(3-2x-E)$

Since $E$ is not zero, then $(3-2x-E)$ must be zero. As $E$ is made smaller and smaller (thus incidentally making the approximation in equation (i) even closer) so $(3-2x-E)$ approaches $3-2x$ as its limit. Equating this to zero gives

$$3-2x = 0$$

or $$x = 1\tfrac{1}{2}$$

It is not clear whether Fermat actually regarded $(3-2x-E)$ as approaching a limit, or whether he regarded $E$ as actually vanishing; but even if the method was not logically rigorous there is no doubt that it worked and that it helped in the development of the calculus.

It will be noted that $(3-2x)$ which is the limiting value of $(3-2x-E)$ is in fact the derivative of $3x-x^2$ which was stated to be the maximum area and that, for this to be so, then $(3-2x)$, or the derivative, must be zero. This is precisely what the differential calculus says.

## MAXIMUM PRODUCT

The same problem, posed as a purely numerical one, is to divide any number into two parts so that the product of the parts shall be a maximum. Let the original number be $n$, and let one of the parts be $x$. Then the other part must be $n-x$, and the product of the parts will be $x(n-x)$.

Then $$y = xn-x^2$$

$$\therefore \frac{dy}{dx} = n-2x = 0 \qquad \text{(for a maximum)}$$

whence $$x = \frac{n}{2} \qquad (\text{,, \quad ,,})$$

That is, whatever value we assign to $n$, the maximum product will result from dividing it into two equal parts. If $n = 100$, this equals $(50+50)$; and $50\times 50 = 2500$. No greater product than this may be achieved. (Checking shows that $49\times 51 = 2499$; $48\times 52 = 2496$ and so on.)

*Maximum volume of a postage parcel*

Assume that Post Office regulations provide that no parcel shall be accepted if the length added to the girth of the parcel exceeds 6 feet but that there shall be no other restrictions. If the parcel is cylindrical, what would be the greatest possible volume of a parcel which conforms with the regulations?

The volume of a cylinder is $\pi r^2 H$. The 'length' of the cylinder is $H$. The 'girth' of the cylinder $= 2\pi r$. The length plus girth is therefore $H+2\pi r$ and the maximum value of this measurement is 6 feet. From these data the maximum volume $y$ is

$$y = \pi r^2 H$$
$$= \pi r^2(6-2\pi r) = 6\pi r^2 - 2\pi^2 r^3$$
$$\therefore \frac{dy}{dr} = 12\pi r - 6\pi^2 r^2$$

Equate to zero for a maximum and then divide by $6\pi r$:

$$12\pi r - 6\pi^2 r^2 = 0$$
$$\therefore 2 - \pi r = 0$$
$$\therefore r = \frac{2}{\pi}$$

Substitute this value of $r$ in the original function and simplify:

$$y = \pi r^2(6-2\pi r)$$
$$= \frac{4}{\pi}(6-4) = \frac{8}{\pi}$$

The maximum volume is therefore $\frac{8}{\pi}$ cu. ft. It is interesting to note that this result is achieved without first calculating the actual length and girth, although these are readily calculated from the result.

$$r = \frac{2}{\pi}$$
$$\therefore 2\pi r \text{ (girth)} = 4 \text{ feet}$$

and $$H \text{ (length)} = 6-\text{(girth)} = 2 \text{ feet}$$

In this example we have substituted an equivalent for $H$ (in terms of $r$) in the function $y$. We could, instead, substitute the equivalent of $r$ (in terms of $H$).

$$H+2\pi r = 6$$
$$\therefore r = \frac{6-H}{2\pi}$$

then
$$
\begin{aligned}
y &= \pi r^2 H \\
&= \pi\left(\frac{6-H}{2\pi}\right)^2 H \\
&= \frac{36H-12H^2+H^3}{4\pi}
\end{aligned}
$$

This is required to be a maximum value. As $4\pi$ is a constant, then it is sufficient to find a value of $H$ such that $36H-12H^2+H^3$ is a maximum. Differentiating and equating to zero gives

$$
\begin{aligned}
36-24H+3H^2 &= 0 \\
\therefore\ (H-6)(H-2) &= 0 \\
\therefore\ H &= 2 \text{ or } 6
\end{aligned}
$$

By the terms of the problem the height cannot be 6 feet since then there would be no girth and nobody wants 6 feet of nothing! The length of the cylinder which will give the maximum volume is therefore 2 feet and the measurement of $r$ (girth) may be found by simple substitution in the appropriate equation above.

The Post Office further stipulates that the length of a parcel at its longest length shall not exceed $3\frac{1}{2}$ feet. The interpretation of 'length' here would include:

(i) longitudinal length (that is, $H$), or
(ii) the diameter of its circular cross-section ($= d = 2r$), or
(iii) the diagonal between the two circular faces ($= \sqrt{H^2+d^2}$).

These values are respectively (i) 2 feet, (ii) 1·27 feet, (iii) 2·37 feet and each of these is within the regulation limit.

*Minimum cost*

What is the most economical speed of flying an aeroplane? In other words, at what speed should an aeroplane be flown so that its flying cost shall be a minimum? Flying cost consists of some expenditure which is fixed relative to speed and some expenditure which varies as the speed varies. Let the flying cost ($C$) per hour at a constant speed ($V$ m.p.h.) be given by the function (entirely imaginary):

$$C = \cdot 0015V^3+81{,}000 \quad \text{(cost units per hour)}$$

over a distance of 1000 miles.

The flight will take $\frac{1000}{V}$ hours; the total cost will therefore be

$$y = \frac{1000}{V}(\cdot 0015V^3 + 81{,}000)$$

$$= \frac{81{,}000{,}000}{V} + 1\cdot 5V^2$$

$$\therefore \frac{dy}{dx} = -\frac{81{,}000{,}000}{V^2} + 3V = 0 \quad \text{(for a minimum)}$$

Multiply by $V^2$ and rearrange:

$$3V^3 = 81{,}000{,}000$$

$$\therefore V = 300 \quad \text{(miles per hour)}$$

The second derivative

$$\frac{d^2y}{dx^2} = \frac{162{,}000{,}000}{V^3} + 3$$

This is positive and the value $V = 300$ therefore gives the desired minimum value for $y$.

It will be appreciated that this example makes a number of assumptions which may not be justified in fact. It assumes the possibility of a constant speed and it assumes that it is possible to express the cost per hour as an explicit function of the variable costs. Whether these are valid assumptions need not concern us here; the example merely shows how, if it is possible to make such assumptions, they may be used to provide further information.

## POINTS OF INFLEXION

It is necessary for maximum and minimum points that $\frac{dy}{dx} = 0$ at those points. This is not sufficient to justify a statement that every point where $\frac{dy}{dx} = 0$ is necessarily a turning point. There is more than one kind of curve. Some curves do not have maximum or minimum values.

Fig. 24 shows the curve of $y = x^3$ and also the curves of the first and second derivatives. The curve of $y = x^3$ has no point so placed that the function is increasing before it and decreasing after it (for a maximum) or vice versa (for a minimum). The function is in fact increasing throughout except at the point $x = 0$ where it is momentarily stationary and the value of $y$ at this point is therefore called a *stationary value*. The curve passes through a point where the derivative is zero (the derivative of $x^3$ is $3x^2$ and this is clearly zero when $x = 0$) but this is not a turning point. At a turning point the upward or downward direction of a curve is reversed. The point on the curve of $y = x^3$ where $\frac{dy}{dx} = 0$ is instead called a *point of inflexion*, this being a point on the curve where the curve's bend changes from clockwise to anti-clockwise, or vice versa, without actually reversing its upward or downward direction. It should be noted that although the point of inflexion in this example occurs when $\frac{dy}{dx} = 0$ this is not always so.

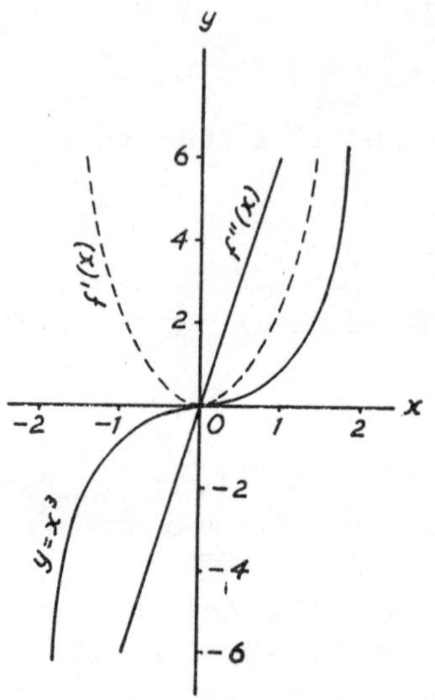


FIG. 24

We have taken this example to attempt to sort out any confusion which can arise when $\frac{dy}{dx}$ *does* equal zero.

The requirement that $\frac{dy}{dx} = 0$ is not therefore sufficient to prove that the point at which this value occurs is necessarily a maximum or a minimum. This means that there is a further problem of making sure that a function value obtained by equating the derivative to zero is in fact either a maximum or minimum before we proceed to check which of these it is. The value might occur at a point of inflexion and not at a turning point at all.

A point of inflexion may be identified as follows. Reference to Fig. 24 shows that the curve of $\frac{dy}{dx}$ has a minimum value at $x = 0$;

that is at the same value of $x$ as at the point of inflexion on the curve of $y$. Thus $\frac{dy}{dx}$ is decreasing before the point of inflexion and increasing after it. Consequently $\frac{d^2y}{dx^2}$ which measures the gradient of the curve of $\frac{dy}{dx}$ is negative before and positive after the point. From this it follows that $\frac{d^2y}{dx^2}$ is zero at the point and this provides the test of identity. In this example the point of inflexion on the curve of $y$ occurs when $\frac{dy}{dx}$ is a minimum. On some curves the point of inflexion occurs when $\frac{dy}{dx}$ is a maximum. In each case however $\frac{d^2y}{dx^2}$ is zero and changing sign at the point.

The tests of discrimination may now be summarized.

*For a maximum point*

$\frac{dy}{dx}$ positive before and negative after the point

$\frac{d^2y}{dx^2}$ negative at the point

*For a minimum point*

$\frac{dy}{dx}$ negative before and positive after the point

$\frac{d^2y}{dx^2}$ positive at the point

*For a point of inflexion*

$\frac{dy}{dx}$ is a maximum or a minimum at the point

$\frac{d^2y}{dx^2}$ is zero and changing sign at the point

The short practical solution to the problem of deciding whether equating $\frac{dy}{dx} = 0$ will identify a turning point is a simple one. If the function value obtained is an impossible one then we may be assured that the function has no turning point and hence no maximum or minimum value. Take the function:

$$y = 2x^3+2x+6$$

Then $$\frac{dy}{dx} = 6x^2+2 = 0 \qquad \text{(for a turning point)}$$

$$\therefore x^2 = -\tfrac{1}{3}$$

and $$x = \pm\sqrt{-\tfrac{1}{3}}$$

This result is impossible. The roots $\pm\sqrt{-\frac{1}{3}}$ are imaginary and there is no real value of $x$ for which $\frac{dy}{dx} = 0$. The curve has no turning point.

The absence of a turning point does not prevent our calculating the greatest possible value of a function within definite limits. The value of $x^3$ continues to increase as $x$ increases and, unless finite limits are imposed, it will continue to increase. Every problem must be considered in the perspective provided by its own restrictions. If we ask what is the greatest possible value of $x^3$ we can only say that this is infinite, for whatever value we substitute for $x$ we can always substitute an even greater value. If however we ask what is the greatest value of $x^3$ such that $x$ is positive and does not exceed the value 6 then, as 6 is the greatest permitted value of $x$, then $6^3 = 216$ is the greatest value of $x^3$ within the prescribed limits.

Every problem which demands a finite answer must set up restrictions upon the function from which the result is to be derived. The calculus will not help us to find the result if there is no turning point, but the truth is that we do not need it if the function is continuously increasing. The greatest value of the function may be calculated by equating $x$ to the value of the upper limit imposed by the problem. This is achieved by a simple numerical substitution for $x$. We do not need the technique of the calculus to crack so small a nut.

## *Chapter 8*

# DIFFERENTIALS

It was noted in Chapter 4 that the symbol $\frac{dy}{dx}$, representing the derivative of a function, originates as a single symbol to represent a single quantity and not as a ratio between $dy$ and $dx$. As yet therefore we cannot regard $\frac{dy}{dx}$ as a fraction of which $dy$ is the numerator and $dx$ is the denominator. This is a logical restriction but it hinders the application of the derivative as a mathematical tool since we cannot regard $dy$ and $dx$ as finite or as having separate meanings of their own.

If the reader glances through the chapters which follow however he will find that $dy$ and $dx$ are very often separated and that $\frac{dy}{dx}$ appears to be treated as if it possesses the very properties of a fraction which we have just noted it cannot be assumed to possess. Thus it might seem that we say one thing but do another. The truth of the matter however is that we are dealing with a *different* $\frac{dy}{dx}$! We have substituted a different two-part symbol in place of the single symbol. Both symbols look exactly the same, which is an unfortunate failing in the notation used, but they have different origins.

This is at first all very confusing and perhaps has a slight smell of sorcery about it. We do not however effect this conversion with a wave of a wand, any more than does a conjuror carry out any real magic. His success depends on his skill of manipulation; our success depends upon showing that although the two symbols are derived from different sources, they are nevertheless numerically equivalent.

We begin with the original non-ratio $\frac{dy}{dx}$. While this does not originate in the form of a fraction involving finite parts, there is no reason why the quantity it represents cannot be put into fractional form. If, for example, the quantity is 2, we can show this as $\frac{4}{2}$ or $\frac{6}{3}$ or some other fraction according to any selected denominator. Each of the fractions is numerically equal to 2.

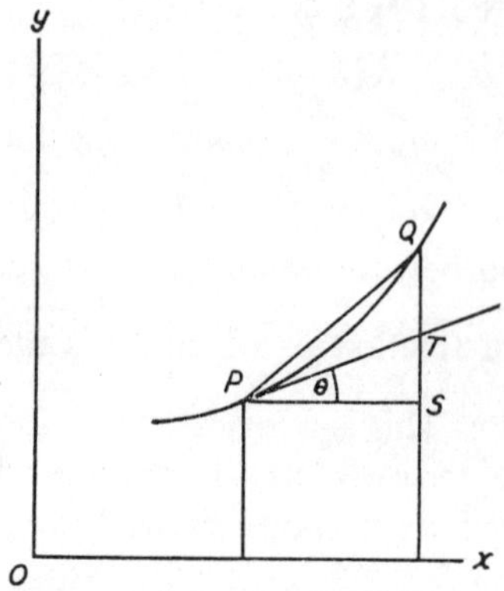


FIG. 25

$\frac{dy}{dx}$ is the limiting value of $\frac{\Delta y}{\Delta x}$ and represents the derivative of $f(x)$. This limiting value is also equal to the gradient of the tangent to the curve. Consequently the derivative of the function and the gradient of the tangent each measures the same quantity. The tangent is a straight line and its gradient may be expressed as a true ratio between quantities, as was shown in Chapter 2. Logically the *derivative of the function* may also be expressed as a true ratio between *quantities related to the tangent*.

This can be most easily understood from a diagram. In Fig. 25 is an arc of the curve $y = f(x)$. $PS$ is drawn parallel to the $x$ axis; $PT$ is the tangent to the curve at $P$. The derivative of the function at $P$ is $\frac{dy}{dx}$. This is measured by the gradient of the tangent, which is equal to $\tan\theta$ or the ratio $\frac{TS}{PS}$.

In relation to the *curve*, $PS = \Delta x$; $QS = \Delta y$. The ratio

$$\frac{QS}{PS} = \frac{\Delta y}{\Delta x}$$

and the limiting value of this is

$$\frac{dy}{dx} \qquad \text{(i)}$$

In relation to the *tangent*, $PS = \delta x$; $TS = \delta y$. The gradient of the tangent is

$$\frac{TS}{PS} = \frac{\delta y}{\delta x} \qquad \text{(ii)}$$

But (i) and (ii) measure the same quantity. Consequently

$$\frac{dy}{dx} = \frac{\delta y}{\delta x}$$

Furthermore the ratio $\frac{\delta y}{\delta x}$ is constant at any point on the tangent and is therefore constant for any value of $\delta x$. If therefore we arbitrarily select a particular value $dx$ and associate this with $dy$ (the related particular value of $\delta y$) then we arrive at

$\frac{dy}{dx}$ or $f'(x)$ the derivative of the function

equals

$\frac{dy}{dx}$ the gradient of the tangent

The particular values $dy$ and $dx$ (of $\delta y$ and $\delta x$ respectively) represent increments in $y$ and $x$ relative to the tangent but are called *differentials* relative to the curve. For the function $y = f(x)$, then $dx$ is called the differential of $x$, and $dy$ is called the differential of $y$.

The derivative is thus equal to the ratio between the differentials and this now explains why the derivative is also called the differential coefficient. In this connection it should perhaps be noted that the originators of the calculus, in describing the derivative as the differential coefficient, were defining the derivative in terms of the differentials. The modern view reverses this assumption and defines the differentials in terms of the derivative.

It will be seen in the diagram that the differential of $x$ is the same as the increment of $x$, since $PS$ is common to the two triangles $PQS$ and $PTS$. This is a purely arbitrary quantity. It would make no difference to the principles involved whatever size we make $dx$. The differential of $y$ however is not the same as the increment of $y$. The value of $dy$ depends upon the choice of value of $dx$. But $\frac{dy}{dx} = f'(x)$ and we therefore define $dy$ as

$$dy = f'(x)\,dx$$

As a result of the equality between the two forms of $\frac{dy}{dx}$ we may in practice regard them as interchangeable. This has a great practical significance since we may separate $dx$ and $dy$ and treat them as finite quantities. The above definition of $dy$ expresses a relationship which figures largely in the development of the integral calculus.

### APPROXIMATE METHODS

Although the differential $dy$ is not equal to $\Delta y$ (the increment in $y$ due to the increment in $x$), nevertheless the difference between these two quantities is practically negligible when they are both very small. In practice therefore when exact results are not insisted upon it is often possible to obtain remarkably good approximations by treating the two quantities as if they are in fact equal.

In this way differentials may be used to calculate the approximate values of increments of functions and their use may be extended to calculations which at first glance do not appear to depend upon the calculus at all.

The reason for using differentials is that their use makes the calculations very much simpler. If we wish to calculate the value of $\Delta y$ it is often much easier to calculate the value of $dy$. The success of this method depends upon $\Delta x$ (and therefore also $\Delta y$ and $dy$) being very small in relation to $x$. This will be evident from Fig. 25; as $\Delta x$ increases so does the difference between $\Delta y$ and $dy$ increase.

The element of error in this method can be very small indeed. Let $y = x^2$. What is the value of $\Delta y$, the increment in $y$, relative to $\Delta x$, the increment in $x$, when $x = 10$ and $\Delta x = 0{\cdot}003$? The increment $\Delta y$ is calculated as follows:

$$y = x^2$$
$$y+\Delta y = (x+\Delta x)^2 = x^2+2x\,\Delta x+(\Delta x)^2$$
$$\therefore \Delta y = 2x\,\Delta x+(\Delta x)^2$$

The differential $dy$ is calculated:

$$y = x^2$$
$$\therefore \frac{dy}{dx} = 2x; \qquad \text{and} \qquad dy = 2x\,dx$$

but $$dx = \Delta x$$
$$\therefore dy = 2x\,\Delta x$$

The *element of error* in regarding $\Delta y$ and $dy$ as being equal is therefore

$$\begin{aligned}\Delta y - dy &= 2x\,\Delta x + (\Delta x)^2 - 2x\,\Delta x\\ &= (\Delta x)^2\\ &= 0{\cdot}000009 \qquad (\text{when } \Delta x = 0{\cdot}003)\end{aligned}$$

The exact value for each quantity is

$$\Delta y = 0{\cdot}060009$$
$$dy = 0{\cdot}06$$

This device helps to perform other calculations which in their original form do not appear to call for the calculation of increments or differentials. If $x = 1{\cdot}998$, what is the value of $x^3 - 2x^2 + 1$? This appears to call for a simple substitution of the value $x = 1{\cdot}998$ but it would be tedious to do this in full. The use of differentials makes it much easier. For this method we regard the value of $x = 1{\cdot}998$ as representing a *change* from $x = 2$ by the addition of $-0{\cdot}002$ to the original value. Then we have $x = 2$; and $\Delta x = dx = -0{\cdot}002$.

Let $$y = x^3 - 2x^2 + 1$$

then $$\frac{dy}{dx} = 3x^2 - 4x$$

$$\begin{aligned}dy &= (3x^2 - 4x)\,dx\\ &= (12-8)(-0{\cdot}002) = -0{\cdot}008\end{aligned}$$

By substituting $x = 2$ in the original equation

$$y = 8 - 8 + 1 = 1$$

But $dy = -0{\cdot}008$. Therefore

$$y + dy = 1 - 0{\cdot}008 = \underline{0{\cdot}992}$$

This is the exact value of $y$ after the addition of the differential $dy$ and the approximate value of $y$ after the addition of the increment $\Delta y$ resulting from the addition of $\Delta x$ to $x$. That is, it is the approximate value of $y$ when $x = 1{\cdot}998$.

The exact value of $y$ when $x = 1{\cdot}998$ calculated the hard way is

$$y = (1{\cdot}998)^3 - 2(1{\cdot}998)^2 + 1 = \underline{0{\cdot}992015992}$$

The approximate answer is as good as most people would want.

## SQUARE ROOTS

Also of interest is a method for calculating the approximate values of those square roots which in any event can never be calculated exactly. This method was known to the ancient Greeks although it was not of course then recognized as a specific application of a general principle.

If the side of a square is $x$, then the area $y = x^2$. If the side of the square is increased to $x+\Delta x$, then the area $y+\Delta y$ increases to $(x+\Delta x)^2$. The increment of area resulting is then

$$\begin{aligned}\Delta y &= (x+\Delta x)^2-x^2\\ &= 2x\,\Delta x+(\Delta x)^2\end{aligned}$$

It was shown (page 47) that if we ignore $(\Delta x)^2$ when $\Delta x$ is very small relative to $x$ we introduce only a negligible element of error. Thus instead of solving

$$(x+\Delta x)^2-x^2 = 2x\,\Delta x+(\Delta x)^2$$

we solve $$(x+\Delta x)^2-x^2 \simeq 2x\,\Delta x$$

whence $$\Delta x \simeq \frac{(x+\Delta x)^2-x^2}{2x} \qquad \text{(A)}$$

This relationship may now be used to calculate the value of $\sqrt{2}$. This value obviously lies between 1 and 2. Since $\sqrt{1{\cdot}96} = 1{\cdot}4$, then we also know that $\sqrt{2}$ must be slightly greater than 1·4. We therefore take $x = 1{\cdot}4$ as our starting value and apply an increment $\Delta x$ so that $x+\Delta x = \sqrt{2}$. Then

$$\begin{aligned}1{\cdot}4+\Delta x &= \sqrt{2}\\ \therefore\ (1{\cdot}4+\Delta x)^2 &= 2\end{aligned}$$

From formula A above

$$\Delta x \simeq \frac{(1{\cdot}4+\Delta x)^2-(1{\cdot}4)^2}{2(1{\cdot}4)}$$

but $$(1{\cdot}4+\Delta x)^2 = 2$$

$$\therefore\ \Delta x \simeq \frac{2-(1{\cdot}4)^2}{2{\cdot}8} = 0{\cdot}014 \text{ approx.}$$

We now have

$$x+\Delta x = \sqrt{2}$$
$$x = 1{\cdot}4$$
$$\Delta x \simeq 0{\cdot}014$$
$$\therefore \sqrt{2} \simeq \underline{1{\cdot}414}$$

If a closer approximation is required the process may be repeated by taking $x = 1{\cdot}414$ as the starting value and applying an increment to this value. This process can be continued indefinitely although in practice no more than one or two repetitions would be required.

This method was originally thought to involve increments only. But in solving the *approximate* equation involving increments

$$(x+\Delta x)^2 - x^2 \simeq 2x\,\Delta x$$

we are in effect really solving the *exact* equation involving differentials

$$(x+dx)^2 - x^2 = 2x\,dx$$

(it being remembered that $\Delta x$ is in fact equal to $dx$).

*Chapter 9*

# BASIC INTEGRATION

The essential technical difference between integration and differentiation is that the latter proceeds by breaking a problem down into little bits and studying the particular value of one or more of these, whereas integration may be regarded as the inverse process of integrating all the little bits into one whole. This is a very loose and unscientific distinction. The real difference can only become apparent as we proceed.

Naturally we do not differentiate a function merely in order that we may then follow the process of integration to get back where we started. While the two processes may be regarded as complementary inversions of each other, they have their separate and distinct purposes.

If you buy a return railway ticket then the outward portion must be used for the outward journey. Which part of the ticket you use depends upon which direction you are travelling. You do not usually travel from A to B merely so that you can travel from B to A. But when, having arrived at B and having completed the object of the journey, you return to A you must use the return half of the ticket. So it is with calculus; which of the two processes you use depends upon what requires to be done.

The ordinary dictionary meaning of the verb '*to integrate*' is 'to combine parts into a whole'. It has much the same general basic meaning in mathematics, although the parts and the whole into which they are combined have special properties. When we integrate an expression we do in effect add a number of parts together, and the whole into which we build them is generally (but not always) the area or volume of a curvilinear figure or solid.

The integral calculus is thus largely concerned with the measurement of shapes, but it also has extremely important applications which do not at first appear to be related to problems of shape. As will be seen later, it is possible to transform many other problems into shape problems, and so to apply the principles of the integral calculus to them.

A great deal of the original thought on which the integral calculus is based is due to Kepler, the discoverer of the laws of planetary motion. Although he was very much concerned with the problems of the universe, Kepler's first contribution to what was eventually to become the integral calculus arose out of a much more down-to-earth need. It is perhaps a surprising thought, but none the less an oddly satisfying one, that the integral calculus which was to open up vast vistas beyond the horizon of man should have had its origin in the simple desire to measure the amount of wine in a cask!

Before describing the principles of the integral calculus it is as well to stress what integration is not. In geometry it is often

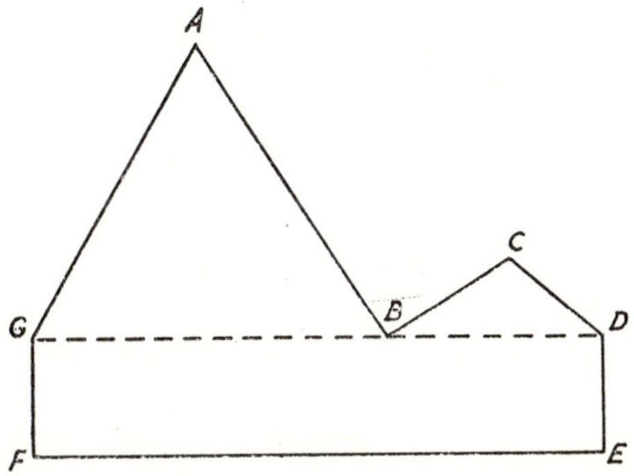


FIG. 26

possible to evaluate the area of an irregular figure by cutting it up into regular or simpler parts and summing their separate areas. Thus the area of the irregular polygon in Fig. 26 might be calculated by summing the areas of the rectangle and the two triangles ABG and CDB. This however does not constitute integration within the special meaning of the calculus. Integration consists of adding together a very large number of very small parts and then taking the limiting value of the sum as the size of the parts is made progressively smaller.

The principles are best introduced by reference to a rectilinear figure. Fig. 27 shows the linear curve $y = x+1$. The $y$ ordinate has been drawn at $x = 10$. This cuts the curve at the point B and effectively encloses the trapezium ABCO. The area of this trapezium is to be measured.

The trapezium is first divided into ten strips of equal width by drawing the ordinates at $x = 1, 2, 3$, etc. For simplicity of presentation only two of these ordinates (at $x = 2$ and 3 respectively) are included in the diagram. These are PN and QM which effectively form one trapezium strip PQMN. The lines SQ and PR are drawn parallel to the $x$ axis, and the ordinate PN is extended to meet SQ at S. This is all the construction required.

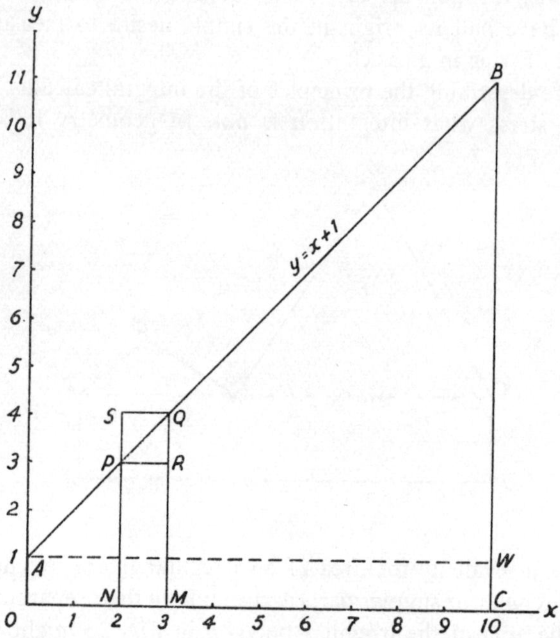


FIG. 27

It may at once be seen that PQMN is smaller than the rectangle SQMN but greater than PRMN. Similarly the area of each trapezium like PQMN (but not illustrated) is less than the area of a larger rectangle like SQMN and greater than the area of a smaller rectangle like PRMN. Consequently the area of the trapezium ABCO, which is equal to the sum of the areas of all the trapezium strips like PQMN, is less than the sum of the areas of all the larger rectangles but greater than the sum of the areas of all the smaller rectangles. This is illustrated in Fig. 28*a* and *b*.

Fig. 28*a* shows the smaller rectangles. In the smallest rectangle the height is 1 unit. The height of the next rectangle is 2 units and the height of each successive rectangle is 1 unit greater than that of the previous one. The width of every rectangle is 1 unit, and the total area of all the rectangles may be calculated by multiplying the sum of their heights by the common width. The total area is therefore 55 square units.

Fig. 28*b* shows the larger rectangles. The height of the first rectangle is 2 units and the height increases by one unit for each

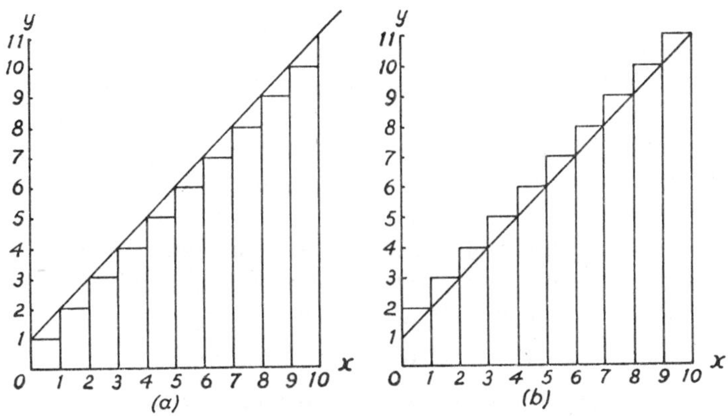


FIG. 28

successive rectangle. The width of 1 unit is again common to all rectangles and their total area is therefore 65 square units.

Accordingly the area ($A$) of the trapezium ABCO is greater than 55 square units but less than 65 square units. That is

$$55 < A < 65$$

If however instead of dividing the trapezium into ten strips, we divide it into one hundred strips, then each strip would be much smaller in width than NM. Similarly the square like SQRP would be much smaller and the difference between the larger and the smaller rectangles relative to any strip would also be much less. As a result of this we would be able to obtain a much better approximation to the value of $A$. The result obtained in this way would actually be

$$59{\cdot}5 < A < 60{\cdot}5$$

Similarly, by using 1000 strips, we would obtain the result

$$59{\cdot}95 < A < 60{\cdot}05$$

and, in general, if we use $n$ strips we will obtain

$$\left(60-\frac{50}{n}\right) < A < \left(60+\frac{50}{n}\right)$$

As we make $n$ greater so $\left(60-\frac{50}{n}\right)$ approaches the value 60 as its limiting value. Similarly, as $n$ is made greater, so also does $\left(60+\frac{50}{n}\right)$ approach the value 60 as its limiting value. $A$ is 'sandwiched' between these two series and the above relationship can hold good only if $A$ provides the limit to both series. Consequently the value of $A$ must be 60 exactly. This can of course be checked from the original diagram (Fig. 27) and it will now be clear why a rectilinear figure was chosen for the demonstration – we may check our result by other methods. The area of the trapezium ABCO is equal to the area of the rectangle AWCO plus the area of the triangle ABW, thus

$$\text{AWCO} + \text{ABW} = \text{ABCO}$$

becomes $(1\times10) + \frac{1}{2}(10\times10) = 60$ square units

This method of calculating an area by showing that it cannot either be greater or less than a particular value was used in antiquity and is known as the method of exhaustion. The difference between the total area of one set of rectangles and the area of the trapezium is diminished, or gradually exhausted, as we make $n$ greater and greater. It has been suggested that in fact this was the beginning of the integral calculus and that Archimedes, who used it, was the real founder of the calculus. But there is no evidence to suggest that Archimedes in using the method had any concept of the 'particular' value as a limiting value of an infinite series. He realized that the true value of the area could not be greater or less than the particular value but he produced no rigorous mathematical formulation as to why this should be so. It is probably nearer the truth to say that Archimedes' method provided the first and all-important clue to the calculus but that, in employing his method, he did not realize its full implication.

We have seen that the area of the trapezium in Fig. 27 could have been calculated by ordinary geometrical calculations without the assistance of the calculus. When we wish to calculate the areas of curvilinear figures however geometry is not sufficient. The principles which were applied to the trapezium associated with

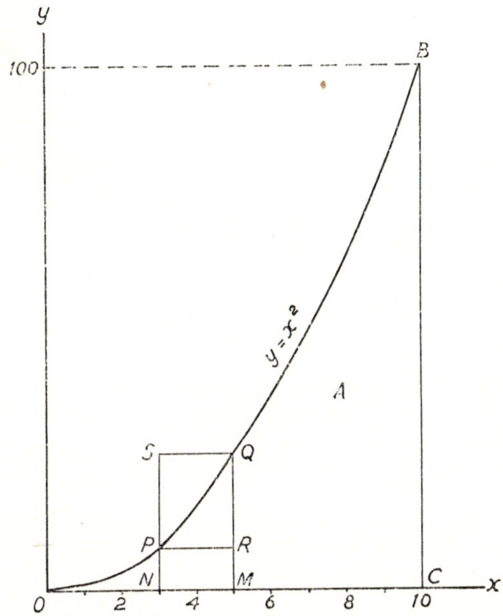


FIG. 29

the linear curve of $y = 1+x$ must now be applied to curvilinear functions.

Fig. 29 shows the curve of the function $y = x^2$. By drawing the ordinate at $x = 10$, we obtain the shape 'under the curve' between the ordinates for $x = 0$ and $x = 10$. It is the area ($A$) of this shape which we propose to measure.

If we divide the area under the curve in Fig. 29 into ten strips and construct rectangles like PRMN and SQMN we can obtain the inequalities

$$285 < A < 385$$

The total area of all the smaller rectangles is 285 square units. The total area of all the greater rectangles is 385 square units.

The area under the curve is between these two values. As we make $n$ (the number of strips) as great as we please, the total of each set of rectangles may be made to approach as close as we like to $333\frac{1}{3}$ square units as a limiting value. Thus the area $A$ is exactly equal to $333\frac{1}{3}$ square units.

There is perhaps one particular point which should be mentioned here. In Fig. 27, the curve of $y = 1+x$ is a straight line and the small rectangles like SQRP are in fact divided into two equal parts by the line of the curve. This is always true even as the rectangles are made smaller. It is fairly easy to understand intuitively that the limiting value is always halfway between the two related approximations. We obtained three separate sets of inequalities for different numbers of strips:

$$55 < A < 65$$

$$59{\cdot}5 < A < 60{\cdot}5$$

$$59{\cdot}95 < A < 60{\cdot}05$$

and $A$ (= 60) is exactly halfway between 55 and 65, or between 59·5 and 60·5, or between 59·95 and 60·05.

In Fig. 29 however the curve of $y = x^2$ is curvilinear and the arc PQ is not a straight line. The area of shape SQP is slightly greater than that of shape PQR, but, as the number of strips, $n$, is made greater, so does the arc PQ become smaller. As it becomes smaller and smaller so does it the more closely approximate to a straight line; so the more closely do the areas of shapes like SQP and PQR approximate to equality with each other, and so does the curvilinear shape PQMN approach the form of the trapezium PQMN (formed by drawing a line between P and Q).

The principles just described now need to be translated into calculus notation. The first thing to note is that the area ($A$) under the curve is itself a function of $x$. It varies as $x$ varies. From Fig. 29 it is clear that the area depends upon the position of the point B on the curve; for each point on the curve, the relative area under the curve is of course different.

The position of point B depends upon the value of $x$ and on the value of $y$. (In Fig. 29 the point B has the co-ordinates $x = 10$; $y = 100$ only because we have made it so; we could have chosen any other point on the curve for our purpose.) But the value of $y$ itself depends upon the value of $x$. Consequently the value of $A$

depends only on $x$. If therefore we knew the exact nature of the functional relationship between $A$ and $x$ then, by differentiating, we could calculate the derivative of $A$ with respect to $x$. That is, we could calculate $\frac{dA}{dx}$. Since we do not know the relationship between $A$ and $x$, we cannot calculate the derivative. On the other hand, if we knew the value of the derivative, we could calculate the value of $A$ by reversing the process of differentiation. This is exactly what the integral calculus enables us to do.

In the discussion of the differential calculus, the symbols $\Delta x$ and $\Delta y$ were used to represent increments or 'little bits' of $x$ and $y$ respectively. These symbols have the same meaning in the integral calculus and, in addition, the symbol $\Delta A$ is now introduced to represent an increment of area ($A$). In Fig. 29, let P be any point on the curve relative to the value of $x$ at N. Then PN $= y$. Let $x$ increase to $x+\Delta x$ at the point M, so that NM $= \Delta x$. But as $x$ increases to $x+\Delta x$, so $y$ increases to $y+\Delta y$ at the point Q by the addition of $\Delta y$, equal to QR. At the same time the area ($A$) increases to $A+\Delta A$, where $\Delta A$ is represented by the curvilinear shape PQMN. (This of course is not the same as the *trapezium* PQMN, but as we have already noted above the latter may be regarded as the limiting form of the former.)

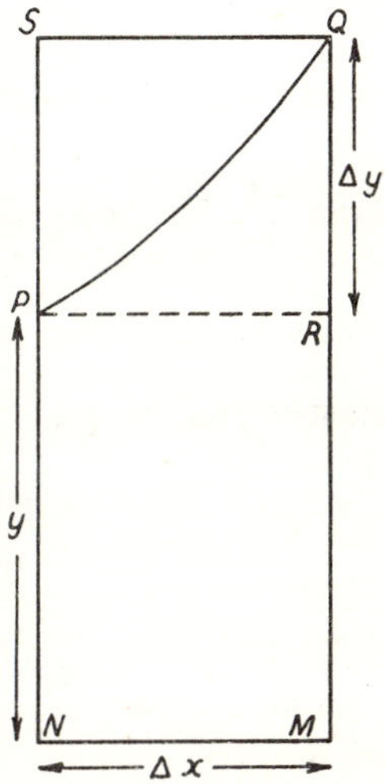


FIG. 30

We now concentrate our attention upon the rectangles SQMN and PRMN. These are reproduced in Fig. 30, from which it can be seen that:

$$\text{Area SQMN} = (y+\Delta y)(\Delta x)$$
$$\text{,, PRMN} = y(\Delta x)$$

The area of the shape PQMN is represented by $\Delta A$. This is smaller than the area of SQMN but greater than the area of PRMN. Consequently we have the inequalities:

$$y\,\Delta x < \Delta A < (y+\Delta y)\,\Delta x \qquad \text{(i)}$$

From this it follows that (dividing throughout by $\Delta x$)

$$y < \frac{\Delta A}{\Delta x} < y + \Delta y \qquad \text{(ii)}$$

Now let $\Delta x$ be decreased indefinitely. As $\Delta x$ approaches zero, so $\Delta y$ approaches zero and, accordingly, $y+\Delta y$ approaches $y$ as its limiting value. Inequality (ii) above means that $\frac{\Delta A}{\Delta x}$ lies between $y$ and $y+\Delta y$. The limiting value of $\frac{\Delta A}{\Delta x}$ is represented in the special form $\frac{dA}{dx}$. The 'inequality' would then disappear. The ratio $\frac{dA}{dx}$ is shown to be neither greater nor less than $y$; this means that $\frac{dA}{dx}$ must be exactly equal to $y$. But $y = x^2$ (from the original function). Hence:

$$\frac{dA}{dx} = x^2$$

In the quantity $\frac{dA}{dx}$ we may recognize the derivative of $A$ when differentiated with respect to $x$. This derivative is now shown as being equal to $x^2$ and this in turn is recognized as the function of $x$ represented by the curve under which the area is to be measured. Accordingly the area $A$ is a function of $x$ such that its first derivative is the same as the function of $x$ represented by the curve. If then we know the value of the derivative of $A$ – as indeed we do from the curve – we may identify the function which represents the area.

Thus if $$\frac{dA}{dx} = x^2$$

then $$A = \tfrac{1}{3}x^3$$

We know this to be true because differentiating the function $A = \frac{1}{3}x^3$ by basic principles reveals the derivative $\frac{dA}{dx} = x^2$.

## RECTANGLE OR TRAPEZIUM?

One important point should be mentioned here. In the preceding pages we have seen that the area under a curve is equal to the limiting value of the sum of the related set of rectangles as the number of these rectangles is indefinitely increased. It may therefore seem contradictory to say that we should not regard the geometrical shape under the curve as necessarily consisting of an indefinite number of rectangles. This is not the same thing at all.

To regard the area as actually consisting of rectangles 'in the limit' cloaks one important stage in the development of the theory. The elements of area (like PQMN in Fig. 27) should rather be regarded as approximations to trapeziums and not to rectangles. In stepping from the finite to the infinite we should really say that the sum of the areas of the trapeziums approaches a limiting value which is equal to the limiting value of the sum of the areas of the related rectangles, and not that each trapezium approaches *identity in all respects* with a rectangle.

We used the concept of rectangles as a convenient method of calculation and also because this method demonstrates the 'trapping' of the area under the curve between the areas of the sets of larger and smaller rectangles. But because the limiting value of the area of a rectangle is equal to the limiting value of the area of its related trapezium, it does not necessarily follow that the trapezium *is* a rectangle any more than a square is a circle merely because it has the same area.

This point need not delay us here, but it will assume importance in dealing with other problems, such as the surface areas of solids.

## RECONSTRUCTING THE CURVE

Although we have been considering integration as essentially a summation process, we have also now seen that it may be effected mathematically by reversing the process of differentiation. It has been shown that the differentiation of a function may be geometrically interpreted as finding the gradient of the curve of the function. As integration may now be regarded as the reverse of differentiation then, since we know the gradient of the curve, we should be able to reconstruct the curve of the function.

This is in fact possible, and is now demonstrated in relation to the function $y = 2x$. This is a linear function so that its curve is a straight line and its gradient $\left(\text{that is, the derivative } \frac{dy}{dx}\right)$ is constantly equal to 2 at every point. This means that the differential triangles drawn in relation to any two points on the curve have the common property of identical proportion between height and base. In this instance $\frac{dy}{dx} = 2$. Therefore $\frac{\Delta y}{\Delta x} = 2$ for every value of $\Delta x$. $\Delta y$ is always twice as great as $\Delta x$.

The function is $y = 2x$. If $x = 0$ therefore then $y = 0$. This means that the curve must cut both the $x$ and $y$ axes at their point of intersection. This is point O in Fig. 31. We select an arbitrary length for $\Delta x$. Mark this length along the $x$ axis so that $OA = \Delta x$. Draw BA parallel to the $y$ axis so that $BA = \Delta y = 2\Delta x$. Join OB to form the first triangle. This procedure is repeated. Draw BC and CD, respectively equal to $\Delta x$ and $\Delta y$. Join BD to form a second triangle.

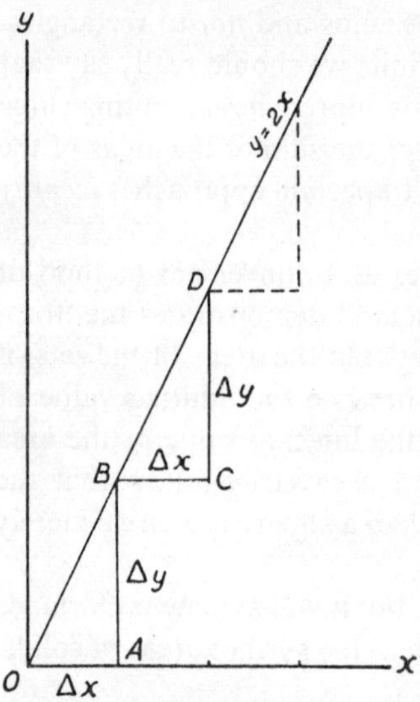


FIG. 31

The two triangles OBA and BDC are similar in every respect. The angle DCB is a right-angle as also is the angle ABC. The angle BDC $= 90° - \text{CBD}$. As the triangles are similar, then the angle OBA $=$ BDC $= 90° - \text{CBD}$.

$$\begin{aligned}\text{Angle OBD} &= \text{Angles OBA} + \text{ABC} + \text{CBD}\\ &= 90° - \text{CBD} + 90° + \text{CBD}\\ &= 180°\end{aligned}$$

OBD is therefore a straight line. The hypotenuse from each of the two triangles together form a straight line which between O and D has the correct gradient of the function $y = 2x$. As it has been shown that one of the co-ordinates lies on this line (at $x = 0$; $y = 0$) and as this line has the same gradient as the curve of the function then this line must also actually form part of that curve. Similar triangles may be constructed in the same way beyond D. It will be seen that this result is achieved irrespective of the size of $\Delta x$. The size of $\Delta x$ is unimportant; it is the ratio between $\Delta x$ and $\Delta y$ which is so important.

Accordingly it will make no difference if we make $\Delta x$ indefinitely small. We may therefore mentally reconstruct the curve from the ratio $\frac{dy}{dx}$ even though we cannot actually draw it on paper. The theoretical result is the same. Applying the same principle to non-linear functions we may construct differential triangles in much the same way. But $\frac{dy}{dx}$ is not constant throughout the curve since it is itself a function of $x$.

For the function $y = x^2$, for example, the derivative $\frac{dy}{dx} = 2x$. Even if we take $\Delta x$ as the same in each triangle, the triangles will not be similar because the ratio between base and height varies for each value of $x$. The 'curve' so drawn will not be a real curve at all. It will merely consist of a number of small straight lines joined together; but it will approximate to the true curve. As we make $\Delta x$ smaller, so we would obtain an even closer approximation until finally in theory, the line would become a smooth curve.

## INTEGRATION METHODS

Having discovered that integration involves the reversal of the differentiation process, we now need to know how to achieve this reversal. The basic rules for finding the derivative of a function are demonstrated in Chapter 5. To find the original function, given the derivative, we merely reverse the action of those rules. It was shown that the derivative $\frac{dy}{dx}$ of the general function $y = ax^n$ is equivalent to $nax^{n-1}$, where $a$ is a numerical coefficient of $x^n$. This applies irrespective of what $y$ represents. It could represent area. We have used the symbol $A$ to represent area but this makes no difference to the properties of the function. $A$ and $y$ are only symbols and for the purposes of mathematics it does not matter what they represent. What does matter is the behaviour of the quantities they represent relative to $x$. Thus in each of the functions

$$y = ax^n$$
$$A = ax^n$$

the relation between the dependent and the independent variables is exactly the same. It is really the same function but employing a different symbol for the dependent variable. It follows therefore that the derivative of $A$ – that is $\frac{dA}{dx}$ – is also equal to $nax^{n-1}$.

To differentiate a function, we proceed as follows:

(i) deduct 1 from the exponent of $x$ (thus $x^n$ becomes $x^{n-1}$)
(ii) multiply the coefficient $a$ by $n$ (thus $a$ becomes $na$)
(iii) as a result of this, $ax^n$ becomes $nax^{n-1}$ (= the derivative).

To find the original function to which the derivative 'belongs' we reverse the procedure and treat the *derivative* as follows:

(i) add 1 to the exponent of $x$ (thus $x^{n-1}$ becomes $x^n$)
(ii) divide the coefficient $na$ by $n$ (thus $na$ becomes $a$)
(iii) as a result of this, $nax^{n-1}$ becomes $ax^n$ (= the original function).

Using this rule we may now find the original function $A = ax^n$ from which the derivative $\frac{dA}{dx} = x^2$ is derived. We proceed as follows (it being noted that the numerical coefficient of $x^2$ is 1).

(i) add 1 to the exponent of $x^2$ to find the value of $n$ (that is $2+1 = 3$)
(ii) divide the coefficient 1 by $n$ to find the value of $a$ (that is $1 \div 3 = \frac{1}{3}$).

Then the original function $A = ax^n$

becomes $$A = \tfrac{1}{3}x^3$$

$A$ represents the area under the curve of $y = x^2$ and accordingly we can now say that this area is equal to $\frac{1}{3}x^3$ square units exactly. This is what we set out to find. It was noted in connection with Fig. 29 that the area of the curve enclosed by the ordinate at $x = 10$ was equal to $333\frac{1}{3}$ square units. This result may now be obtained from the formula. Here $x = 10$; and the area is therefore

$$A = \tfrac{1}{3}(10)^3 = 333\tfrac{1}{3}$$

Thus far we have been concerned only with the area under the curve of the function $y = x^2$. The principle involved however is a general one and applies (with certain restrictions which will be referred to later) to all functions of $x$. It has been shown in this chapter that $\frac{dA}{dx}$ must be exactly equal to $y$, and this is so no matter what function of this general form is represented by $y$. This may be understood if reference is again made to Fig. 30 and to the argument which was developed to show that $\frac{dA}{dx} = y$. Everything that was then said applies with equal justification to the curves of any other functions of the form $y = ax^n$ which we could have drawn instead of the curve of $y = x^2$.

It was noted in connection with this latter example that:

$$\Delta A \simeq y\,\Delta x \qquad \text{[See page 107; equation (i)]}$$

$\Delta A$ is the increment in area relative to $\Delta x$ the increment in $x$. It is therefore the area of a strip like PQMN in Fig. 29, and the area of this strip is approximately equal to $y\,\Delta x$ (the area of a rectangle $y$ by $\Delta x$), the approximation being as close as we like to make it by taking $\Delta x$ sufficiently small. This applies to all strips of area like PQMN. The sum of a finite number of these strips is therefore approximately equal to the sum of the related rectangles. That is

$$\sum \Delta A \simeq \sum y\,\Delta x$$

the approximation again depending on the smallness of $\Delta x$. As the number of strips is increased (that is, as $\Delta x$ is decreased) so the sum of the areas of the rectangles approaches the sum of the area of the strips. The sum of the area of the strips is the area under the curve. Accordingly the area under the curve is the limiting value of the sums of the areas of the rectangles. This relationship is expressed

$$A = \int dA = \int y\,dx$$

the symbol $\Delta x$ being replaced by $dx$ to indicate that we are dealing with a limiting value. The area under the curve is thus 'integrated' by summing a number of values and taking the limiting value of their sum as the number of values is indefinitely increased. This idea of summation and then passing to the limit of the sum is the basic idea of the integral calculus and is involved or implied in every application of it.

In order to calculate the area from the formula obtained we have to integrate the expression $y\,dx$; that is we have to sum all the values like $y\,\Delta x$ and take the limiting value of this sum as $\Delta x$ is indefinitely decreased. But we have seen that to calculate the area we merely reverse the process of differentiation. Thus when we have to integrate an expression we merely pick out that part of it which is the derivative of the original function so as to identify the latter.

The mathematical process of integration is the inverse of differentiation. Differentiation involves the finding of a derivative; integration involves the finding of the original function once we know its derivative. Thus the instruction $\int x^2\,dx$, which we would originally have treated as meaning 'find the limiting value of the sum of all values like $x^2\,dx$' may instead be interpreted as the instruction 'find the function of which the derivative is $x^2$'. This means exactly the same as the sentence: 'If $\frac{dA}{dx} = x^2$, find the value of $A$' but it says it in a concise notation.

In $\int x^2\,dx$, the expression which follows the $\int$ sign (that is $x^2\,dx$) is called the *integrand*. The symbol $dx$ is included in this as an indication that the expression to which it is attached represents a derivative. It has no numerical value in the expression. In fact the $\int$ and $dx$ symbols should be regarded as inseparable. The expression $\int x^2$, for example, is meaningless on its own; it requires the inclusion of the $dx$ symbol, as in $\int x^2\,dx$, to indicate that $x^2$ is the derivative of the function (which is to be identified) on being differentiated with respect to $x$.

The original function may be described in either of two ways, depending upon which particular aspect of it is stressed. Regarded as the original function giving rise to the derivative which forms part of the integrand, it is called the *prime* or *primitive* of the latter. Regarded merely as the result of integrating the integrand it is called the *indefinite integral.* (The significance of the word *indefinite* will become evident later in this chapter.)

One of the useful facts about integration is that once we have integrated an expression and discovered the primitive, we may then check our working by differentiating the latter once more. The derivative thus obtained should of course be the same as that with which we started in the original integrand. This check is a simple one and should never be omitted.

So far we have assumed that the original function which is to be found by integrating is of the form $ax^n$. Not all functions are of this simple form however and it is necessary to consider certain complications which may arise. It will be remembered (page 56) that additive constants in a function are not carried through to the derivative. The derivative of $y = x^2$ is $2x$. But the derivative of $x^2+1$ or, more generally, of $x^2+c$ (where $c$ is any constant) is also $2x$. Thus

$$\text{if } y = x^2 \qquad \text{then } \frac{dy}{dx} = 2x$$

$$\text{,, } y = x^2+c \qquad \text{,, } \frac{dy}{dx} = 2x$$

When each of these functions is differentiated, the constant is ignored since it does not contribute to the rate of growth of the function. When we come to integrate however and are told that the derivative is $2x$, we are faced with the problem that this derivative could have been derived from any one of these functions. That is to say that we cannot assume that the true function is $y = x^2$; we have no idea what constant $c$ may have been ignored in the process of differentiating. It is always safer to assume that there was a constant in the original function and accordingly say that

$$\int 2x\,dx = x^2+c$$

The value of $c$ may be zero, when $\int 2x\,dx$ would in fact be equal to $x^2$, but we are not entitled to *assume* that this is so.

From this it is evident that we shall not be able to formulate a set of rules for the integration of each and every function so as always to obtain a definite numerical value. In the differential calculus we can formulate a set of general rules, but the very practical rules, which make differentiating relatively simple, themselves serve to make integrating that much more difficult or even impossible. We ignore additive constants in differentiating and, if we do not know what these constants were, this may well baffle us when we attempt to integrate.

But we may be able to calculate the value of the constant from observed values. If, for example, we find on integrating that the function is $y = x^2+c$ then, if we also know the value of $y$ relative to just one value of $x$, we may determine the value of $c$. If $y = 10$ when $x = 3$, then

$$y = x^2+c$$

becomes
$$10 = 3^2+c$$

whence
$$c = 1$$

This fact comes to our assistance when measuring the area under a curve which cuts the $y$ axis as in Fig. 29. The area under this curve was shown as $\frac{1}{3}x^3$ square units. In finding this area by integration we now see that it should have been shown as $\frac{1}{3}x^3+c$. The area is therefore really

$$A = \tfrac{1}{3}x^3+c$$

Reference to the diagram however shows that when $x = 0$, so also does $A = 0$.

That is:

$$0 = 0+c$$

whence $c$ must also be zero at that point. But $c$ is constant and is therefore zero for every value of $x$.

Hence when we measure the area under this curve between the $y$ axis and any ordinate, the constant which appears in the integral may be ignored. It must be made clear however that we do not ignore the *existence* of the constant. We ignore it only because its value is zero. This is an important reservation since in many applications of the integral calculus, the value of $c$ is not zero. In any particular calculations we must first satisfy ourselves that $c$ is zero before we may ignore it.

The expression $x^2+c$ is called the indefinite integral of $2x$ because it contains the undetermined constant $c$ which, in its turn, is described as the constant of integration. Some confusion

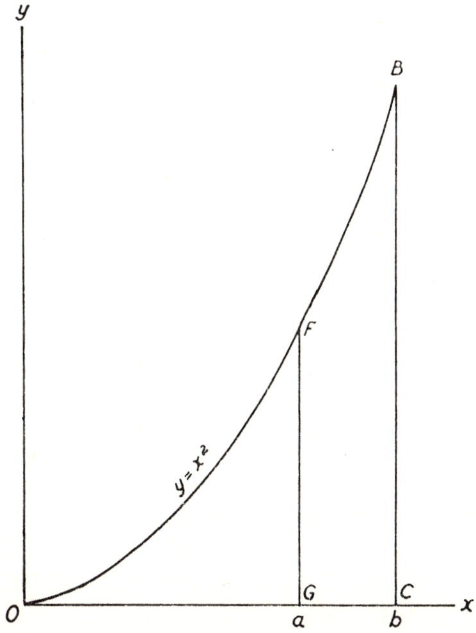


FIG. 32

exists in the use of the word *integral*. Sometimes it is used for the whole expression including the integral sign $\left(\text{such as } \int 2x\,dx\right)$ *before* the expression is integrated. Sometimes it is used for the result (such as $x^2+c$) *after* the expression has been integrated.

Having identified the indefinite integral, we now proceed to the definite integral. In our example of the calculation of the area under the curve of the function $y = x^2$ we were measuring the area between the $y$ axis and the ordinate drawn at a specific value of $x$ (that is, at point C in Fig. 29). It is often desirable however to calculate the area between two ordinates as, for example, between FG and BC in Fig. 32. This diagram again shows the curve of $y = x^2$.

The ordinate FG is drawn at $x = a$ and the area of the shape OFG may be designated as $A_a$. The ordinate BC is drawn at $x = b$ and the area of the shape OBC may be designated as $A_b$. From our previous calculations we now know that $A_a$ and $A_b$ may each be calculated independently from the formula for the indefinite integral.

$$A = \int x^2\,dx = \tfrac{1}{3}x^3 + c$$

All that has to be done in each instance is to substitute the value of $x$ at which the ordinate is drawn.

The area $A_b$ of shape OBC is therefore $\frac{1}{3}b^3+c$, since $x = b$ at BC. Similarly the area $A_a$ of shape OFG is $\frac{1}{3}a^3+c$, since $x = a$ at FG. The area which is to be measured however is that of shape FBCG between FG and BC and the diagram shows that this may be calculated as the difference between $A_b$ and $A_a$. That is

$$\begin{aligned}\text{Area FBCG} &= (\tfrac{1}{3}b^3+c)-(\tfrac{1}{3}a^3+c)\\ &= \tfrac{1}{3}b^3-\tfrac{1}{3}a^3\end{aligned}$$

and the area between the ordinates is thus determined.

This process is known as integrating within limits; that is within the limits of application formed by the respective ordinates. It should be clearly understood that here the words 'within limits' are used in the sense of defining the boundaries of the area to be calculated. This must not be confused with the meaning of 'limits' of infinite series.

The notation of the indefinite integral $\int x^2\,dx$ or its equivalent $\frac{1}{3}x^3+c$ represents the integral for any value of $x$. It is indefinite because of the intrusion of the unknown constant $c$ and also because it does not specify the relative value or values of $x$ from which the integral's numerical equivalent may be calculated. It will be seen however that when we integrate within limits the constant $c$ very conveniently disappears. It originally appears in each of the integrals representing the areas $A_b$ and $A_a$, but when $A_a$ is deducted from $A_b$, the constant is cancelled out. The resulting integral, which is the one which has been calculated within limits (and which in the present instance is $\frac{1}{3}b^3-\frac{1}{3}a^3$), is called the *definite integral.* It no longer involves the constant $c$ but depends solely upon the values of $x$ represented by $a$ and $b$.

The notation for the definite integral in our example is

$$\int_a^b x^2\,dx$$

It will be seen that this is basically the same as the indefinite integral but that in addition it shows the limits within which integration is to be performed. This is the form of the definite integral before integration has been performed. After the integration has been performed, the definite integral is shown as

$$\left[\frac{1}{3}x^3\right]_a^b$$

This means 'substitute $a$ for $x$ in the expression within the brackets and deduct the value thus obtained from the value obtained by substituting $b$ for $x$'. That is, 'deduct $\frac{1}{3}a^3$ from $\frac{1}{3}b^3$'.

Suppose therefore that we wish to calculate the area under the curve of $y = x^2$ between the ordinates at $x = 3$ and $x = 10$. This area is

$$\int_3^{10} x^2\,dx$$

$$= \left[\frac{1}{3}x^3\right]_3^{10}$$

$$= \tfrac{1}{3}(10)^3 - \tfrac{1}{3}(3)^3 = 324\tfrac{1}{3} \text{ square units}$$

This method may now be applied to a problem with which we are familiar. What is the distance fallen by a stone within a specified period of time, if the speed at which the stone is falling is known?

In Chapter 5 it was shown that this speed is continuously increasing throughout the fall as the time ($x$) increases; but that it is always equal to $32x$. We are now dealing with speed (in feet per second) represented by $y$ as a function of $x$ (time in seconds). This is shown in Fig. 33 by the linear curve of $y = 32x$. The distance fallen is then represented by the area under the curve.*

To calculate the total distance fallen by the stone in the first

* Page 221.

three seconds we draw the ordinate at $x = 3$ and measure the area under the curve (that is, the area of the triangle ABO). Using the formula for the area under a curve we have

$$y = 32x$$

$$\therefore A = \int 32x\,dx \qquad \text{(indefinite integral)}$$

$$= \int_0^3 32x\,dx \qquad \text{(definite integral within limits } x = 0;\ x = 3\text{)}$$

$$= \Big[16x^2\Big]_0^3$$

$$= 16(3)^2 - 0 = 144 \text{ feet}$$

This result is the same as was shown on page 10.

## FUNDAMENTAL THEOREM

The fact that the definite integral is indeed really a limiting value as $\Delta x$ approaches zero (between the limits $x = a$ and $x = b$) is expressed

$$\lim_{\Delta x \to 0} \sum_{x=a}^{x'=b} f(x)\,\Delta x = \int_a^b f(x)\,dx$$

(where the symbol $\sum$ represents 'the sum of a finite number of terms like . . .') and the integral sign may be regarded as an abbreviation of the limit notation on the left side of the equation.

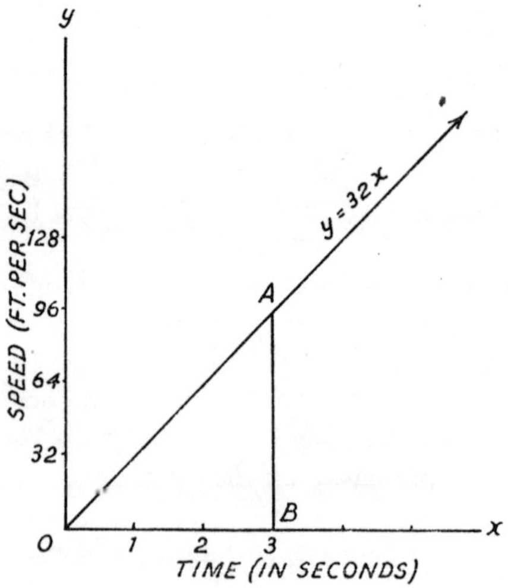


FIG. 33

Integration is essentially a process of summation and is defined as such above. It may nevertheless be effected by reversing the process of differentiation, and the fact that the integral

$$F(x) = \int f(x)\,dx$$

itself has a derivative which is equal to $f(x)$, that is

$$F'(x) = f(x)$$

is commonly called the fundamental theorem of the calculus.

*Successive integration*

In Chapter 5 it was shown that the first derivative of a function may itself be differentiated to give a second derivative and that this process may be continued until a derivative is reached which is not itself a function of $x$. This is called successive differentiation. It may be reversed by successive integration.

If we know that the second derivative $\frac{d^2y}{dy^2}$ of a function $y$ is $5x$, we can proceed to discover the original function by successive integration. If we integrate $\int 5x\,dx$, this will give as the first derivative

$$\frac{5x^2}{2}+c_1$$

We then integrate $\int\left(\frac{5x^2}{2}+c_1\right)dx$ to obtain the original function. Thus

$$y=\frac{5x^3}{6}+c_1x+c_2$$

The notation for successive integration is a simple development of the notation already introduced. If we are to integrate twice then we merely use the $\int$ and $dx$ symbols twice. The indefinite integral would then be shown as

$$\int\int 5x\,dx\,dx$$

This means 'integrate $5x$ twice with respect to $x$'. If we were required to integrate three times (that is, if the expression is the third derivative of a function) this is shown as

$$\int\int\int 5x\,dx\,dx\,dx$$

It will be noted that reversing the process of differentiation will introduce a different constant at each stage. These are distinguished as $c_1$ and $c_2$ above. If however we apply limits for each stage of integration then the constants will disappear just as conveniently as they do in one-stage integration.

As an example we now perform successive integrations of

$$\int_0^3 \int_1^2 5x \, dx \, dx$$

*First stage:* Integrate $\int 5x \, dx$ between the limits $x = 1$ and $x = 2$.

$$\int 5x \, dx = \frac{5x^2}{2} + c$$

$$\therefore \int_1^2 5x \, dx = \left[\frac{5x^2}{2}\right]_1^2 = \frac{15}{2}$$

*Second stage:* Integrate $\int \frac{15}{2} dx$ between the limits $x = 0$ and $x = 3$.

$$\int \frac{15}{2} dx = \frac{15x}{2} + c$$

$$\therefore \int_0^3 \frac{15}{2} dx = \left[\frac{15x}{2}\right]_0^3 = 22\tfrac{1}{2}$$

The result, $22\frac{1}{2}$, is the numerical value of the original function for the given limits.

# Chapter 10

# SUBSTITUTION METHODS

The previous chapters have dealt with basic ideas and principles and have shown how the differential calculus and the integral calculus are inter-related. In this chapter we revert to the differential calculus and consider some technical points so as to build up our ability to differentiate a wide variety of functions.

Prominent among calculus methods is the method of substitution which helps to simplify many technical processes. If we have a complicated-looking function it often helps to change its appearance. Thus if $y = (x^2+2x)^5$ we may instead write this as

$$y = u^5$$

where $u = (x^2+2x)$. The expression $(x^2+2x)$ is a function of $x$ and in substituting $u$ for it we have merely changed its name. We have not changed the character of the function itself.

This introduces the idea of a *function of a function.* This simply means that, in certain circumstances, one quantity may be expressed as a function of another quantity which is itself a function of a third quantity. Thus $(x^2+2x)$ is a function of $x$. Similarly $(x^2+2x)^5$ is a function of $(x^2+2x)$.

To differentiate a function of a function, two separate stages are necessary. If $y = (x^2+2x)^5$ then the first stage is to differentiate $y$ with respect to $(x^2+2x)$. A further stage is then required to find the derivative of $y$ with respect to $x$. The whole is greatly simplified by ordinary substitution.

Let $\qquad y = (x+2x)^5$

substitute $\qquad u = x^2+2x$

whence $\qquad y = u^5$

This may easily be differentiated by the ordinary method. (This differentiation is with respect to $u$, not with respect to $x$.)

$$\frac{dy}{du} = 5u^4$$

This measures the rate of growth of $y$ relative to the growth of $u$. To find the rate of growth of $y$ relative to $x$ we take note of the relationship

$$\frac{dy}{dx}=\frac{dy}{du}\times\frac{du}{dx}$$

which may be obtained as follows.

Let $\Delta y$, $\Delta x$, and $\Delta u$ be increments in $y$, $x$, and $u$ respectively. These are finite quantities. Then by the properties of ordinary fractions

$$\frac{\Delta y}{\Delta x}=\frac{\Delta y}{\Delta u}\times\frac{\Delta u}{\Delta x}$$

since on the right-hand side the symbols $\Delta u$ cancel out. As $\Delta x$ approaches zero, each of the three fractions approaches a limit; the limiting value of $\frac{\Delta y}{\Delta x}$ will then be equal to the product of the other two limiting values. That is, if $y=f(u)$ where $u=f(x)$ then

$$\frac{dy}{dx}=\frac{dy}{du}\times\frac{du}{dx}$$

The significance of this requires emphasis. $y$ is a function of $u$. The derivative $\frac{dy}{du}$ represents the rate at which $y$ increases with respect to $u$. In its turn, $u$ is a function of $x$. The derivative $\frac{du}{dx}$ represents the rate at which $u$ increases with respect to $x$. Then the rate at which $y$ increases with respect to $x$ is equal to the product of these two separate rates. That is, if $y$ increases 3 times as fast as $u$, and $u$ increases 4 times as fast as $x$, then $y$ increases 12 times as fast as $x$.

In our example we have substituted

$$u=x^2+2x$$

therefore $$\frac{du}{dx}=2x+2$$

It has already been shown that $\frac{dy}{du} = 5u^4$. Consequently

$$\frac{dy}{dx} = \frac{dy}{du} \times \frac{du}{dx} = (5u^4)(2x+2)$$

and, as $u = x^2+2x$, then

$$\frac{dy}{dx} = 5(x^2+2x)^4 \times (2x+2)$$
$$= 10(x+1)(x^2+2x)^4$$

## DIFFERENTIATION OF A PRODUCT

A function may be expressed as the product of two factors, as in

$$y = (x^4+x^3)(x^2+c)$$

where $c$ is a constant. There are two ways in which functions of this type may be differentiated. In simple cases, the easiest method is to multiply out and differentiate the separate terms. That is

$$y = x^6+x^5+cx^4+cx^3$$

then
$$\frac{dy}{dx} = 6x^5+5x^4+4cx^3+3cx^2$$

From this result it will be apparent that the derivatives of the factors cannot be multiplied to give the final derivative. If such a course were possible then, the derivatives of $x^2+c$ being $2x$, the effect of the constant $c$ would not be represented in the solution. The constant, which is additive only in the factor $x^2+c$, becomes a multiplying constant in the function considered as a whole.

The second method of differentiating a product provides a general rule for all products. For simplicity the function of this form may be expressed as

$$y = uv$$

where $u$ and $v$ are different functions of $x$. (In the example given above, $u$ would equal $(x^4+x^3)$ and $v$ would equal $(x^2+c)$.) As $x$ grows to $(x+\Delta x)$, so $y$, $u$, and $v$ grow respectively to $(y+\Delta y)$; $(u+\Delta u)$; and $(v+\Delta v)$. Then

$$y+\Delta y = (u+\Delta u)(v+\Delta v)$$
$$= uv+u\,\Delta v+v\,\Delta u+\Delta u\,\Delta v$$

but
$$y = uv$$

so
$$\Delta y = u\,\Delta v+v\,\Delta u+\Delta u\,\Delta v$$

and, dividing throughout by $\Delta x$, then

$$\frac{\Delta y}{\Delta x} = u\frac{\Delta v}{\Delta x} + v\frac{\Delta u}{\Delta x} + \Delta u\frac{\Delta v}{\Delta x}$$

The limiting values of $\frac{\Delta y}{\Delta x}$; $\frac{\Delta v}{\Delta x}$; and $\frac{\Delta u}{\Delta x}$ are $\frac{dy}{dx}$; $\frac{dv}{dx}$; and $\frac{du}{dx}$ respectively. At the same time, as $\Delta x$ approaches zero, so does $\Delta u$ approach zero, and consequently so also does $\Delta u\frac{\Delta v}{\Delta x}$ approach zero. Hence, the limiting value:

$$\frac{dy}{dx} = u\frac{dv}{dx} + v\frac{du}{dx}$$

The derivative of the function $y = uv$ is therefore found by adding together:

(i) the product of: $u$ and the derivative of $v$
(ii) the product of: $v$ " " " $u$

In the example $y = (x^4+x^3)(x^2+c)$ we have

$$u = x^4+x^3 \qquad \text{whence } \frac{du}{dx} = 4x^3+3x^2$$

$$v = x^2+c \qquad \text{whence } \frac{dv}{dx} = 2x$$

Then

$$\begin{aligned}\frac{dy}{dx} &= u\frac{dv}{dx} + v\frac{du}{dx} \\ &= (x^4+x^3)2x+(x^2+c)(4x^3+3x^2) \\ &= 2x^5+2x^4+4x^5+4cx^3+3x^4+3cx^2 \\ &= 6x^5+5x^4+4cx^3+3cx^2\end{aligned}$$

This is the same result as was obtained in the first method by multiplying the factors and then differentiating each term of the product. In this example the second method has no great advantage over the first method. The important point is that the second method provides a formula for the differentiation of all products including those which cannot easily be differentiated by the first method.

## DIFFERENTIATION OF A QUOTIENT

A function may be expressed as the quotient of two expressions; that is, the result of dividing one expression by another. As division by $x$ may be regarded as multiplication by $\frac{1}{x}$, it is logical that a method of differentiating a quotient may be derived from the method of differentiating a product.

In order to differentiate $y = \frac{1}{x}$ it is first to be noted that $\frac{1}{x}$ may also be represented as $x^{-1}$. This can be differentiated quite simply by the basic rule for $y = ax^n$ where $n = -1$; and $a = 1$. Multiply the coefficient of $x$, by the exponent number; and deduct 1 from the exponent thus giving the derivative of $x^{-1}$ as $-1x^{-2}$ or, more simply, $-x^{-2}$ or $-\frac{1}{x^2}$.

These ideas are now applied to the differentiation of the function

$$y = \frac{x+1}{x^2}$$

This may be expressed differently as

$$y = (x+1)(x^{-2})$$

or

$$y = uv$$

where $u = x+1$; $v = x^{-2}$. Then $\frac{du}{dx} = 1$; $\frac{dv}{dx} = -2x^{-3}$. From the product rule

$$\begin{aligned}\frac{dy}{dx} &= u\frac{dv}{dx}+v\frac{du}{dx}\\ &= (x+1)(-2x^{-3})+(x^{-2})(1)\\ &= -2(x+1)(x^{-3})+x^{-2}\\ &= \frac{-2(x+1)}{x^3}+\frac{1}{x^2}\end{aligned}$$

which may be further simplified to $-\left(\frac{x+2}{x^3}\right)$.

In practice however this method tends to become complicated and it is usual to employ another method derived from it.

Let $$y = \frac{u}{v}$$

whence $y = uv^{-1}$ or $uw$ where $w = v^{-1}$. Then by the product rule:

$$\frac{dy}{dx} = u\frac{dw}{dx} + w\frac{du}{dx}$$

The two terms on the right-hand side of the equation are now considered separately:

*First term:* $u\frac{dw}{dx}$

Since $w = v^{-1}$ then $w$ is a function of $v$ and may be differentiated with respect to it in the normal way:

$$\frac{dw}{dv} = -v^{-2}$$

Then, multiplying both sides of this equation by $\frac{dv}{dx}$,

$$\frac{dw}{dv} \times \frac{dv}{dx} = (-v^{-2})\frac{dv}{dx}$$

and, simplifying and substituting the equivalent of $-v^{-2} = -\frac{1}{v^2}$ on the right-hand side

$$\frac{dw}{dx} = \left(-\frac{1}{v^2}\right)\frac{dv}{dx}$$

whence $$u\frac{dw}{dx} = -\left(\frac{u}{v^2}\right)\frac{dv}{dx}$$

*Second term:* $w\frac{du}{dx}$

Since $w = v^{-1}$ then

$$w\frac{du}{dx} = (v^{-1})\frac{du}{dx}$$

$$= \left(\frac{1}{v}\right)\frac{du}{dx}$$

*Substitution in equation for* $\frac{dy}{dx}$

The equation $$\frac{dy}{dx} = u\frac{dw}{dx} + w\frac{du}{dx}$$

becomes $$\frac{dy}{dx} = -\left(\frac{u}{v^2}\right)\frac{dv}{dx} + \left(\frac{1}{v}\right)\frac{du}{dx}$$

$$= \frac{-u\dfrac{dv}{dx}}{v^2} + \frac{v\dfrac{du}{dx}}{v^2}$$

$$= \frac{v\dfrac{du}{dx} - u\dfrac{dv}{dx}}{v^2}$$

If therefore $$y = \frac{x+1}{x^2}$$

let $u = x+1$; $v = x^2$; whence

$$\frac{dy}{dx} = \frac{v\dfrac{du}{dx} - u\dfrac{dv}{dx}}{v^2}$$

$$= \frac{(x^2)(1)-(x+1)(2x)}{(x^2)^2} = -\left(\frac{x+2}{x^3}\right)$$

thus agreeing with the result already obtained.

The rule for the differentiation of a quotient is therefore: multiply the denominator by the derivative of the numerator; then multiply the numerator by the derivative of the denominator. Subtract the second product from the first and divide the difference by the square of the denominator.

## DIFFERENTIATION OF A ROOT

To differentiate a function which is expressed as a root of a function of $x$, we may proceed as follows:

Let $$y = \sqrt{1+x^2}$$

This may be rewritten as

$$y = (1+x^2)^{\frac{1}{2}}$$

Let $\quad u = 1+x^2 \quad \left(\text{So } \frac{du}{dx} = 2x\right)$

Then $\quad y = u^{\frac{1}{2}} \quad \left(\text{So } \frac{dy}{du} = \frac{1}{2}u^{-\frac{1}{2}}\right)$

and
$$\frac{dy}{dx} = \frac{dy}{du}\times\frac{du}{dx}$$
$$= (\tfrac{1}{2}u^{-\frac{1}{2}})2x$$
$$= xu^{-\frac{1}{2}} = x(1+x^2)^{-\frac{1}{2}}$$

or
$$\frac{x}{\sqrt{1+x^2}}$$

Square roots often occur in problems involving maximum or minimum values and the ability to differentiate them enables us to find solutions to the problems. Here is an example (however improbable the circumstances!). An escaping convict has to cross a river from point A on one bank to reach his home (B) 6 miles along the river on the opposite bank. With the good fortune normally attendant upon characters in mathematical problems he has the use of a rowing boat and a bicycle. He therefore rows to a point on the other bank and then cycles home along the towpath. His average speeds are 8 m.p.h. rowing and 10 m.p.h. cycling. The river is 3 miles wide and its banks are parallel. What is the quickest time in which he can reach home and to what point on the opposite bank must he row to achieve this?

The reader is advised to draw a diagram consisting of a rectangle with A at the bottom left corner and B at the top right corner. This rectangle represents the stretch of the river 6 miles long and 3 miles wide. Let Z be the point to which he rows and let the distance rowed be $x$ miles. This will be represented by drawing the line AZ; this in turn is the hypotenuse of a right-angled triangle of which one of the other sides is 3 miles. Then the third side (which coincides with part of the other bank) $= \sqrt{x^2-9}$.

Z is therefore $\sqrt{x^2-9}$ miles away from a point on the bank directly opposite A; it is therefore $6-\sqrt{x^2-9}$ miles from B. The distances travelled are therefore $x$ miles at 8 m.p.h. and $6-\sqrt{x^2-9}$ miles at 10 m.p.h. The total time to cover the journey is therefore:

$$t = \frac{x}{8}+\frac{6-\sqrt{x^2-9}}{10} = \frac{x}{8}+\frac{3}{5}-\frac{\sqrt{x^2-9}}{10} \quad \text{(hours)}$$

The next step is to calculate the derivative and equate it to zero for a minimum (see Chapter 7). First however we must be able to differentiate $\dfrac{\sqrt{x^2-9}}{10}$ by substitution.

Let $x^2-9 = u$ $\left(\text{whence } \dfrac{du}{dx} = 2x\right)$; then $y = \dfrac{u^{\frac{1}{2}}}{10}$ $\left(\text{whence } \dfrac{dy}{du} = \dfrac{u^{-\frac{1}{2}}}{20}\right)$. Then

$$\frac{dy}{dx} = 2x\left(\frac{u^{-\frac{1}{2}}}{20}\right)$$

$$= \left(\frac{2x}{20}\right)\left(\frac{1}{\sqrt{x^2-9}}\right) = \frac{x}{10\sqrt{x^2-9}}$$

We can now differentiate the function

$$t = \frac{x}{8}+\frac{3}{5}-\frac{\sqrt{x^2-9}}{10}$$

$$\frac{dt}{dx} = \frac{1}{8}+0-\frac{x}{10\sqrt{x^2-9}} = 0 \qquad \text{(for a minimum)}$$

Simplifying and transposing gives:

$$\sqrt{x^2-9} = \frac{4x}{5}$$

$$\therefore\ x = 5 \quad \text{(miles = distance by boat)}$$

Then $$6-\sqrt{x^2-9} = 6-\sqrt{16}$$
$$= 2 \quad \text{(miles = distance cycled)}$$

That is, Z is 2 miles along the bank from B. The time taken on the whole journey is then found by substituting for $x$ in the function:

$$t = \frac{5}{8}+\frac{3}{5}-\frac{\sqrt{16}}{10} \quad \text{(hours)}$$

$$= \frac{33}{40} \quad \text{(hours – or } 49\tfrac{1}{2} \text{ minutes).}$$

The total distance covered is 7 miles; this is not in fact the shortest route (the diagonal from A to B is about 6·7 miles) but it is the quickest route in the particular circumstances. The long way round is often the quickest!

Substitution or illumination?



Substitution is a very useful device which helps us to make complicated expressions more amenable. Other kinds of transformation are also possible. The differentiation of a fraction, for instance, may be made easier if the fraction is first represented in a different form. One way of doing this is by the use of partial fractions whereby a complicated fraction is first represented as the sum of simpler fractions. If

$$y = \frac{3x+1}{x^2-1}$$

this fraction may be split into partial fractions as

$$\frac{3x+1}{x^2-1} = \frac{1}{x+1} + \frac{2}{x-1}$$

$$= (x+1)^{-1} + 2(x-1)^{-1}$$

and the derivative may be found as follows, first substituting

$$u = x+1 \qquad \text{whence } \frac{du}{dx} = 1$$

$$v = x-1 \qquad \text{whence } \frac{dv}{dx} = 1$$

Then $\qquad y = u^{-1} + 2v^{-1}$

and the two terms may be differentiated separately.

*First term* $u^{-1}$

$$\frac{dy}{du} = -u^{-2}$$

$$\therefore \frac{dy}{dx} = \frac{dy}{du} \times \frac{du}{dx} = (-u^{-2})(1) = -(x+1)^{-2}$$

*Second term* $2v^{-1}$

$$\frac{dy}{dv} = -2v^{-2}$$

$$\therefore \frac{dy}{dx} = \frac{dy}{dv} \times \frac{dv}{dx} = (-2v^{-2})(1) = -2(x-1)^{-2}$$

Therefore the total derivative is

$$\frac{dy}{dx} = -(x+1)^{-2} - 2(x-1)^{-2}$$

$$= -\frac{1}{(x+1)^2} - \frac{2}{(x-1)^2} = -\frac{3x^2+2x+3}{(x^2-1)^2}$$

This result can of course be checked by the ordinary quotient rule: Let

$$y = \frac{3x+1}{x^2-1} = \frac{u}{v}$$

$$v = x^2-1 \qquad \therefore \frac{dv}{dx} = 2x$$

$$u = 3x+1 \qquad \therefore \frac{du}{dx} = 3$$

$$\therefore \frac{dy}{dx} = \frac{v\frac{du}{dx} - u\frac{dv}{dx}}{v^2}$$

$$= \frac{3(x^2-1)-(3x+1)2x}{(x^2-1)^2} = -\frac{3x^2+2x+3}{(x^2-1)^2}$$

In this particular example, chosen for its comparative simplicity of demonstration, there is little to choose between the two different methods. But where the original function is a really complicated expression, the method of partial fractions (not to be confused with partial differentiation) is much easier to handle. A note on the method by which a fraction may be represented as a sum of partial fractions is included in Appendix I.

## INVERSE FUNCTIONS

It has been noted that if $y$ may be expressed as a function of $x$ then, in general, $x$ may in turn be expressed as a function of $y$. Thus if

$$y = 4x$$

then

$$x = \frac{y}{4}$$

The second function is called the inverse function of the first.

The derivative of the inverse function will measure the rate of change of $x$ relative to $y$, that is $\frac{dx}{dy}$. The derivative of the original function is $\frac{dy}{dx}$. But

$$\frac{dx}{dy} = \frac{1}{\frac{dy}{dx}} \qquad \text{since } \frac{dy}{dx} \times \frac{dx}{dy} = 1$$

and it follows therefore that the derivative of the inverse function is equal to the reciprocal of the derivative of the original function.

In the above example, therefore:

$$\frac{dy}{dx} = 4$$

so

$$\frac{dx}{dy} = \tfrac{1}{4}$$

The usefulness of this relationship between the original and its inverse function is that, in order to obtain the derivative of the former, it may sometimes be easier to obtain the derivative of the latter first and then merely to calculate its reciprocal.

(1)

(2)

## *Chapter 11*

# EXPONENTS AND LOGARITHMS

Functions do not all grow in the same way. The growth of the function $y = 2x$ is constant, whereas the growth of the function $y = x^3$ is not. We now turn to a particular type of growth which is such that the increment of growth within a given period of time is always proportional to the magnitude of the original quantity as increased by previous increments. As each increment is added so the magnitude of the quantity increases. Successive increments bear a fixed ratio to the quantity as it exists when the increments are added. As the quantity grows larger so also do the increments grow larger.

This kind of growth is usually referred to as organic growth since it describes, subject to certain conditions and limitations, the growth of a living organism. A plant, for instance, grows slowly at first but then its rate of growth increases as the size of the plant increases. A plant will of course stop growing eventually for one reason or another. Other similar types of growth may in theory be continued indefinitely, although there may be practical limitations to their operation.

The growth of money at stake in a double-or-quits process behaves in this way. Brown owes Smith £1. He suggests that they toss a coin. If he calls correctly then he owes Smith nothing; if he calls incorrectly then his debt is doubled. He loses the first toss and so owes Smith £2. He then loses the next toss and then owes Smith £4. If he continues to lose then the amount he owes Smith will grow in accordance with the sequence

$$1 \quad 2 \quad 4 \quad 8 \quad 16 \quad 32 \quad \ldots .$$

each term being twice as great as the next preceding term.

If $y$ is the amount of the debt, then $y$ is doubled after every toss of the coin. The increment of growth at every toss is thus proportional to the value to which $y$ has grown at that time. The rate of increase does not look alarming in the figures shown but if Brown loses 20 consecutive tosses then his debt will have grown to over one million pounds! This is highly improbable, but it

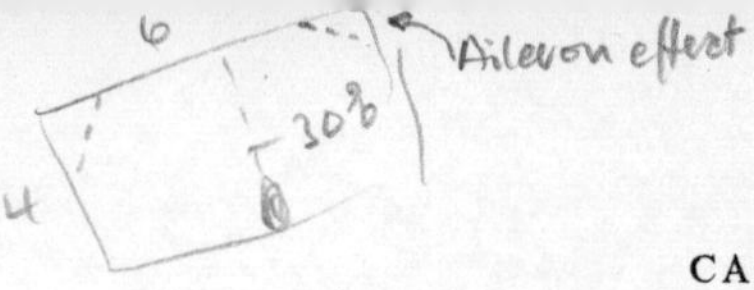




illustrates the remarkable growth potential of a function of this nature.

The value of $y$ jumps from 2 to 4 and from 4 to 8 and so on. There are no intervening values but this is only because of the limitations of the conditions. The number of tosses jumps from 1 to 2; you cannot have $1\frac{1}{2}$ tosses. The amount of the debt depends upon the number of tosses and must therefore jump in the same way. If $x$ be the number of tosses then the value of $y$ may be shown as follows for the respective values of $x$.

$$x = 1; \qquad y = 2 = 2^1$$
$$x = 2; \qquad y = 4 = 2^2$$
$$x = 3; \qquad y = 8 = 2^3$$

In each instance the exponent of 2 in the value of $y$ is identical with the value of $x$ and, generally

$$y = 2^x$$

In this particular instance $x$ is always an integer and $y$ is not a continuous function. Nevertheless there are continuous functions in which $x$ and $y$ are related in this way, and it is necessary to find out how their rates of growth may be measured.

It may be a little difficult to visualize such a function. In the example above it was easy to calculate the amount of the debt relative to any number of tosses. The latter were clearly distinguishable and could be considered separately. A continuous function however changes its value during every interval, no matter how small, even though the change during that interval is so small as to be physically indistinguishable. Its value changes and then almost before we have had time to consider its change it has changed again. Yet the growth in the second interval will in fact bear a fixed ratio to the value of the function as increased by the growth in the first interval. An organism of course grows independently of mathematics. The nature of its growth may be described by mathematics but it does not depend upon it. An organism grows in a particular way; it is up to mathematics to catch up with it.

This subject is best approached through a study of the compound interest type of growth which in practice is not a continuous function, but we may consider what would happen if it were. This may seem an odd idea but this example is invariably used because most people understand the principles of compound interest and may therefore concentrate on the mathematics.

Compound interest growth behaves in the way we have described. If we invest £100 at a compound interest rate of 10% per annum, then at the end of the first year the investment will be worth £110. If this amount is left invested then the interest in the second year will be 10% of £110 and the value at the end of the second year will be £121. The annual growth then is

| | |
|---|---|
| 1st year | £10 0*s*. |
| 2nd year | £11 0*s*. |
| 3rd year | £12 2*s*. |

and continues to increase. By the tenth year the annual interest will have grown to over £23. At simple interest the annual rate would be constant at £10.

Compound interest is usually computed at the end of each year. This suggests that in the first year the original £100 remains at that level throughout the year right up to the last day when its value apparently jumps to £110. But since the money has presumably not been lying idle it is reasonable to suggest that perhaps it should have been earning interest throughout the year. Thus halfway through the year the principal amount of £100 should perhaps have grown to £105, the addition of £5 being equivalent to half a year's interest at 10%. But in this case the second half of the year's interest would have to be calculated on the increased amount invested – that is, on £105. The interest for the second part of the year would then be £5 5*s*. and the total interest for the year would be £10 5*s*. instead of the £10 interest calculated only once at the end of the year.

If $P$ be the principal amount and if $r$ be the rate of interest then the interest, if calculated at the end of the year, will be $P$ times $\frac{r}{100}$. The principal will have grown to $\left(P+\frac{Pr}{100}\right)$ or $P\left(1+\frac{r}{100}\right)$. At the end of the second year, if interest is still calculated annually, the principal will have grown to $P\left(1+\frac{r}{100}\right)^2$ and, generally, after $s$ years it will have grown to $P\left(1+\frac{r}{100}\right)^s$.

Now suppose that interest is calculated half-yearly instead of annually. At the end of the first half-year the principal will have grown to $P\left(1+\frac{r}{200}\right)$, the interest added being one-half of $\frac{r}{100}$. At the end of the second half-year the principal will have grown to $P\left(1+\frac{r}{200}\right)^2$ and at the end of $s$ years to $P\left(1+\frac{r}{200}\right)^{2s}$.

But once we ask why do we not calculate interest half-yearly instead of annually, we may also ask why it should not be calculated monthly. If this were to be done, then the principal would grow to

$$P\left(1+\frac{r}{12\times 100}\right)^{12s}$$

after $s$ years. And then we ask – why not weekly, daily, hourly or instantaneously, or, in general, $t$ times a year. This latter notion provides the formula for the principal after $s$ years as

$$P\left(1+\frac{r}{100t}\right)^{ts}$$

In this expression $\frac{r}{100t}$ is always less than 1. Indeed as $t$ grows larger so $\frac{r}{100t}$ grows smaller since $r$, the nominal rate of interest, is constant.

This expression may be transformed. Let $\frac{r}{100t}=\frac{1}{n}$ whence $t=\frac{nr}{100}$ and $ts=\frac{nrs}{100}$. The expression then becomes

$$P\left(1+\frac{1}{n}\right)^{\frac{nrs}{100}}$$

$$=P\left[\left(1+\frac{1}{n}\right)^n\right]^{\frac{rs}{100}} \qquad \text{(i)}$$

In the special case where the rate of interest is 10% per annum and the amount is allowed to grow for 10 years, then $r$ and $s$ are each equal to 10 and the fraction $\frac{rs}{100}$ simplifies to 1; and the amount of the principal at the end of the tenth year would be

$$P\left(1+\frac{1}{n}\right)^n$$

We now suppose that interest is added at indefinitely small intervals of time since this is a required property if the value of the principal is to be regarded as a continuous function. The value of $t$ will then be indefinitely large as also will the value of $n$ since $t = \frac{nr}{100}$ and, since $r$ is constant, so $n$ varies as $t$ varies. Again, $\frac{r}{100t}$ and therefore $\frac{1}{n}$ will be indefinitely small.

In expression (i) above the values of $P$, $r$, and $s$ are all finite and, in any actual calculation, these values would all be given as basic data. Consequently we are left with the problem of calculating the value of $\left(1+\frac{1}{n}\right)^n$ as $n$ grows indefinitely large. That is, we have to calculate the limiting value of $\left(1+\frac{1}{n}\right)^n$ as $n \rightarrow \infty$ (that is, as $n$ approaches infinity).

We have hitherto dealt with the limit of a function as the independent variable approached zero. We now have to deal with the 'limit' of a function as the independent variable approaches infinity. This might be supposed to introduce impossibilities of calculation. We have seen that as we make the number of interest calculations occur more frequently so the amount of interest increases. Then, by making the number of calculations indefinitely large, do we not also make the amount of interest indefinitely large? That is, does the amount of interest attain almost astronomical proportions and, if this is so, how can we talk of a limiting value of it? We can see that as $n \rightarrow 0$ then 0 is the limiting value of $n$. But as $n \rightarrow \infty$ then we cannot conveniently fix our mind on any finite limit for $n$. The short answer to this question is that, although the amount of interest continues to grow as we make the number of calculations more frequent, the rate of this increase is actually slowing down. The truth of this statement may be demonstrated mathematically as follows:

We are concerned with the limiting value of $\left(1+\frac{1}{n}\right)^n$. As $n \to \infty$; so $\frac{1}{n} \to 0$. Instead of concentrating on $n$ therefore we concentrate on $\frac{1}{n}$ and transform our problem into one which we can handle. For this transformation we rely upon the Binomial Theorem.

The first few terms of the expansion of the binomial $\left(1+\frac{1}{n}\right)^n$ are as follows:

$$\left(1+\frac{1}{n}\right)^n = 1+\frac{n}{1}\cdot\frac{1}{n}+\frac{n(n-1)}{\underline{|2}}\cdot\frac{1}{n^2}+\frac{n(n-1)(n-2)}{\underline{|3}}\cdot\frac{1}{n^3}\ \ldots$$

Each term of the expression may be simplified so as to remove from the denominators the terms involving $n$. Thus

$$\left(1+\frac{1}{n}\right)^n = 1+1+\frac{1-\frac{1}{n}}{\underline{|2}}+\frac{\left(1-\frac{1}{n}\right)\left(1-\frac{2}{n}\right)}{\underline{|3}}+\ \ldots$$

In each of the numerators of the expansion we now regard the fractions like $\frac{1}{n}$ and $\frac{2}{n}$ etc. as being indefinitely small. Then as $\frac{1}{n} \to 0$, so $\left(1-\frac{1}{n}\right)$ and similar expressions such as $\left(1-\frac{2}{n}\right)$ each $\to 1$. The limiting value of the third term of the expansion is therefore $\frac{1}{\underline{|2}}$. Likewise, the limit of the fourth term is $\frac{(1)(1)}{\underline{|3}} = \frac{1}{\underline{|3}}$ and, generally, the limit of the $r$th term is $\frac{1}{\underline{|r}}$.

But the limiting values of the individual terms of the expansion will together equal the limiting value of the binomial represented by the expansion. Therefore the limiting value of $\left(1+\frac{1}{n}\right)^n$ is equal to $$1+1+\frac{1}{|\underline{2}}+\frac{1}{|\underline{3}}+\frac{1}{|\underline{4}}+\ .\ .\ .\ +\frac{1}{|\underline{r}}$$

and the sum of this series is called $e$.

We cannot calculate the value of this limit exactly but, as in the case of $\pi$, we can obtain a value for it which may be as nearly accurate as we like to make it. The first ten terms of the expansion are numerically as follows:

1·
1·
0·5
0·16 666 667
0·04 166 667
0·00 833 333
0·00 138 889
0·00 019 841
0·00 002 480
0·00 000 275

2·71 828 152

The value of $e$, correct to six decimal places, is 2·718282. In this calculation we have taken only ten terms but it will be seen even from these that the value of successive terms falls off rapidly. The eleventh term would be 0·000000275 and it will be seen that the addition of this term has no significant effect upon the first six decimals of the approximate value of 2·718282.

If we now revert to the special case of a principal accumulating at 10% compound interest over 10 years then, if the growth of the capital is continuous, the principal will have reached $P\left(1+\frac{1}{n}\right)^n$ at the end of the ten years. This amount is now known to be $2{\cdot}718282\times P$. That is, £1 would grow to £2·718282.

In the above demonstration the calculations have been based on the assumption that the growth of the principal is continuous in time and that interest is not added at finite intervals. It is well known however that interest is never calculated in this way. In practice, interest is always calculated as a definite percentage increment for a definite period of time. The growth of the principal is not continuous but, as was noted earlier, there are other forms of growth which do behave in this way. The growth of a function in which the growth is always proportional to the function may be measured in the way described, and this has been the object of our demonstration. These functions are normally found in advanced scientific theory and it would be less easy to demonstrate the growth of these without expecting the reader to be fully acquainted with these scientific applications.

The limiting value of $\left(1+\frac{1}{n}\right)^n$ is equal to $e$, as $n$ approaches infinity. Therefore the expression for the value of the principal (shown in expression (i) above) may be rewritten as

$$P(e)^{\frac{rs}{100}}$$

when $n$ is indefinitely large.

This expression is of the form $Pe^x$ $\left(\text{where } x = \frac{rs}{100}\right)$ and is an example of an exponential function. Unlike other functions, this function has a variable exponent while the factor ($e$ in this instance) is constant. Exponential functions occur in a number of scientific calculations and before we can differentiate them we need to know something more about them. We have seen that

$$e = \lim_{n\to\infty}\left(1+\frac{1}{n}\right)^n$$

It follows that

$$e^x = \lim_{n\to\infty}\left(1+\frac{1}{n}\right)^{nx}$$

The binomial $\left(1+\frac{1}{n}\right)^{nx}$ may now be expanded using the Binomial Theorem

$$\left(1+\frac{1}{n}\right)^{nx} = 1+\frac{nx.1}{n}+\frac{nx(nx-1)}{\underline{|2}}\cdot\frac{1}{n^2}+\frac{nx(nx-1)(nx-2)}{\underline{|3}}\cdot\frac{1}{n^3}+\dots$$

or, simplified

$$= 1+x+\frac{x^2-\frac{x}{n}}{\underline{|2}}+\frac{x^3-\frac{3x^2}{n}+\frac{2x}{n^2}}{\underline{|3}}+\dots$$

As $n$ becomes indefinitely great, the fractions involving $n$ in the denominators become indefinitely small. That is as $n \longrightarrow \infty$ so $\left(x^2-\frac{x}{n}\right) \longrightarrow x^2$. Similarly $\left(x^3-\frac{3x^2}{n}+\frac{2x}{n^2}\right) \longrightarrow x^3$. Consequently the third and fourth terms of the expansion approach $\frac{x^2}{\underline{|2}}$ and $\frac{x^3}{\underline{|3}}$ respectively. The limiting value of $\left(1+\frac{1}{n}\right)^{nx}$ is equal to the sum of the limits of the separate terms in the expansion. Thus

$$e^x = \lim_{n\to\infty}\left(1+\frac{1}{n}\right)^{nx} = 1+x+\frac{x^2}{\underline{|2}}+\frac{x^3}{\underline{|3}}+\ldots\frac{x^r}{\underline{|r}}$$

This series is called the exponential series because it is the series for the exponential function. Other related series may be obtained

$$e^{-x} = 1-x+\frac{x^2}{\underline{|2}}-\frac{x^3}{\underline{|3}}+\ldots$$

$$e^{ax} = 1+ax+\frac{a^2x^2}{\underline{|2}}+\frac{a^3x^3}{\underline{|3}}+\ldots$$

by substituting $-x$ and $ax$ respectively in place of $x$ in the orginal series.

The curve of $e^x$ would represent the continuous growth of a function of the type considered in the compound interest demonstration. The increments grow larger as the function increases.

Conversely the curve of $e^{-x}$ would show the continuous decrease of a function in which each decrement is proportional to the changing magnitude of the function. This would illustrate a type of dying-away law such as may be encountered in chemistry and physics. The cooling of a hot body for example is represented in Newton's law of cooling by the equation

$$\theta_t = \theta_0 e^{-at}$$

where $\theta_0$ is the original excess of temperature of a body over the temperature of its surroundings; $\theta_t$ is the remaining excess after a period of time, $t$; and $a$ is the constant of decrement.

To differentiate $e^x$ all that we need do is to differentiate each separate term in the usual way and then find the sum of all the separate differentiations. Thus

$$y = e^x = 1 + x + \frac{x^2}{\underline{|2}} + \frac{x^3}{\underline{|3}} + \frac{x^4}{\underline{|4}} + \ldots$$

$$\therefore \frac{dy}{dx} = 0 + 1 + \frac{2x}{\underline{|2}} + \frac{3x^2}{\underline{|3}} + \frac{4x^3}{\underline{|4}}$$

and this may be simplified to:

$$\frac{dy}{dx} = 1 + \frac{x}{1} + \frac{x^2}{\underline{|2}} + \frac{x^3}{\underline{|3}} + \ldots$$

But this is exactly the same series as for $e^x$ itself! Consequently the derivative of the function $y = e^x$ is also $e^x$. And by reversing the process

$$\int e^x \, dx = e^x + c$$

This may at first seem a remarkable result. Yet it is not so remarkable really. The derivative of a function is a measure of the rate of change of the function, and the function $e^x$ is itself defined as a function which has a rate of change proportional to itself. It will also be found that no matter how often the series for $e^x$ is differentiated, the result – that is, the first, second, third, and every successive derivative – is equal to $e^x$. This is the only function with this peculiar property.

Exponential functions are not always in a form as simple as $y = e^x$. It may be necessary, for example, to differentiate the function $y = e^{4x^3}$. This function is still of the general exponential form but, although the derivative of $e^x$ is also equal to $e^x$, it is not true that the derivative of $e^{4x^3}$ is also equal to $e^{4x^3}$. To differentiate this function it is necessary to resort to the use of substitution.

Let $y = e^{4x^3}$; what is the value of $\frac{dy}{dx}$? The derivative $\frac{dy}{dx}$ is the result of differentiating $y$ with respect to $x$.

If, instead of writing $e^{4x^3}$, we write $e^u$ then the derivative of the function $\left(\text{i.e. } \frac{dy}{du}\right)$ is also $e^u$. This is the result of differentiating $y$ with respect to $u$. We want to find the value of $\frac{dy}{dx}$ and not of $\frac{dy}{du}$

and the method of calculating this is as follows:

$$y = e^{4x^3}$$

Let $\quad 4x^3 = u \quad$ whence $y = e^u$ and $\dfrac{dy}{du} = e^u$

$$\text{also } \frac{du}{dx} = 12x^2$$

But
$$\frac{dy}{dx} = \frac{dy}{du} \times \frac{du}{dx} = e^u \times 12x^2$$
$$= e^{4x^3} \times 12x^2$$

## LOGARITHMS

It has been shown that the amount ($A$) to which the principal ($P$) would grow at a continuous rate of compound interest is of the form

$$A = Pe^x$$

whence
$$\frac{A}{P} = e^x$$

Now the logarithm of a number to a specific base may be defined as the exponent of the power to which the base is to be raised to give the required number. If therefore $e$ is taken as the base, then $x$ will represent the logarithm of $\frac{A}{P}$. More generally, if $y = e^x$ then $\log_e y = x$. Thus $e^2$ is approximately equal to 7·389 and $\log_e (7{\cdot}389) = 2$.

In point of time, logarithms to the base $e$ preceded those to the base 10, for it was to the base $e$ that Napier calculated the original logarithms. For this reason they are sometimes called Napierian logarithms. They are also variously called Hyperbolic or Natural logarithms to distinguish them from common logarithms which are to the base 10. For ordinary arithmetical calculations it is usual to employ common logarithms because of the greater ease with which they may be employed. In the study of physical phenomena however in which the exponential function frequently occurs, Napierian logarithms are used because of the peculiar properties of that function.

A function may be expressed in the form

$$y = \log_e x$$

This relationship may be transformed to

$$x = e^y$$

Thus instead of $y$ being expressed in terms of $x$, we now have $x$ expressed in terms of $y$. This is the inverse function of the original function. We adopt this transformation because we know how to deal with the function $e^y$ whereas we do not as yet know how to deal with the function $\log_e x$.

The derivative of the function $e^y$ is the same as the function itself. But the derivative is of the form $\frac{dx}{dy}$. That is, it is the derivative of $x$ with respect to $y$. The desired derivative is $\frac{dy}{dx}$. We therefore have

$$\frac{dx}{dy} = e^y$$

$$\therefore \frac{dy}{dx} = \frac{1}{e^y}$$

But $$e^y = x$$

$$\therefore \frac{dy}{dx} = \frac{1}{x}$$

Hence the derivative of $y = \log_e x$ is $\frac{1}{x}$ or $x^{-1}$.

It was noted* in Chapter 5 that it was not possible to obtain a derivative of the form $x^{-1}$ by the ordinary rules of differentiation. Now we see that it is possible to find a derivative of this form but that it can be derived only in respect of a particular type of function.

This of course is of immense assistance in integrating. For if $\frac{dy}{dx} = x^{-1}$, then $\int x^{-1}\,dx = \log_e x + c$ by reversing the process. That is

$$y = \log_e x$$

then $$\frac{dy}{dx} = x^{-1}$$

* Page 54.

and of course $$\int x^{-1}\,dx = \log_e x+c$$

This example serves to illustrate that integration truly is the reverse of differentiation. We cannot integrate an integrand to identify the original function unless the integrand has previously been derived from that function by the process of differentiation. Mountaineers crossing a glacier use ropes to mark out their path as they proceed in order that they may be able to retrace their steps later. Integration may be effected only by following the path mapped out by differentiation.

By the same method as that used above it may be shown that if $y = \log_e (x+a)$ then the derivative $\frac{dy}{dx} = \frac{1}{x+a}$ since we merely substitute $(x+a)$ for $x$. And by reversing the process

$$\int \frac{1}{x+a}dx = \log_e (x+a)+c$$

The substitution used here is made possible by the rule for the differentiation of a function of a function. In Chapter 10 it was seen that if $y = f(u)$ where $u = f(x)$, then the derivative $\frac{dy}{dx}$ is obtained as

$$\frac{dy}{dx} = \frac{dy}{du}\times\frac{du}{dx}$$

If then we have the function

$$y = \log_e f(x)$$

where $u = f(x)$, we may differentiate as:

$$\frac{dy}{du} = \frac{1}{f(x)}$$

$$\frac{du}{dx} = f'(x)$$

$$\therefore \quad \frac{dy}{dx} = \frac{1}{f(x)}\times f'(x) = \frac{f'(x)}{f(x)}$$

and, reversing the process on integrating,

$$\int \frac{f'(x)}{f(x)}dx = \log_e f(x)+c$$

This therefore provides a general rule for the integration of a fraction which is in such a form or may be transformed so that the numerator may be recognized as the derivative of the denominator. The integral of such a fraction is equal to the logarithm of the denominator.

A function of the form $y = \log_e x^a$ may be differentiated by using the rule that $\log_e x^a = a \log_e x$.

Thus if $$y = \log_e x^3$$

then, also, $$y = 3 \log_e x$$

and $$\frac{dy}{dx} = 3 \times \frac{1}{x} = \frac{3}{x}$$

This may also be worked out by substituting

$$y = \log_e x^3$$

Let $x^3 = u$ whence (i) $y = \log_e u$ and $\frac{dy}{du} = \frac{1}{u}$

$$\text{(ii)}\ \frac{du}{dx} = 3x^2$$

Then $$\frac{dy}{dx} = \frac{dy}{du} \times \frac{du}{dx} = \frac{1}{u} \times 3x^2 = \frac{3x^2}{x^3} = \frac{3}{x}$$

The function $y = e^x$ is really a special form of exponential function of which the general form is $y = a^x$ where $a$ is any positive number. To differentiate this general form we again have recourse to logarithms. It should be noted that although the derivative of $e^x$ is also $e^x$, the derivative of $a^x$ is not also $a^x$ (except of course in the special case when $a = e$).

The method is as follows, using the logarithm method.

$$y = a^x$$

$$\therefore \log_e y = \log_e a^x = x \log_e a$$

$$\therefore \quad x = \log_e y \times \frac{1}{\log_e a}$$

Now differentiate $x$ with respect to $y$. Note here that the derivative of $\log_e y = \frac{1}{y}$; and that $\frac{1}{\log_e a}$ is constant.

Then $$\frac{dx}{dy} = \frac{1}{y} \times \frac{1}{\log_e a} = \frac{1}{a^x} \times \frac{1}{\log_e a} = \frac{1}{a^x(\log_e a)}$$

Since $\frac{dy}{dx}$ is the reciprocal of $\frac{dx}{dy}$ then

$$\frac{dy}{dx} = a^x \log_e a$$

and, reversing the process,

$$\int a^x \log_e a \, dx = a^x + c$$

The derivative of $a^x$ is therefore equivalent to $a^x$ multiplied by $\log_e a$. In the special case when $a = e$, then

$$\frac{dy}{dx} = e^x \log_e e$$

$\text{Log}_e\, e$ is equal to 1; hence $\frac{dy}{dx} = e^x$.

When $y$ is expressed as a function of $x$ such that the exponent of $a$ is itself a function of $x$, the principle of substitution is again required.

Let $$y = a^{3x^2}$$

and then let $u = 3x^2$ whence (i) $\frac{du}{dx} = 6x$

(ii) $y = a^u$

$$\therefore \log_e y = u \log_e a$$

$$\therefore u = \log_e y \times \frac{1}{\log_e a}$$

$$\therefore \frac{du}{dy} = \frac{1}{y} \times \frac{1}{\log_e a}$$

and $$\frac{dy}{du} = y \log_e a$$

But $$\frac{dy}{dx} = \frac{dy}{du} \times \frac{du}{dx} = y \log_e a \times 6x \quad \text{(A)}$$

$$= a^{3x^2} \log_e a \times 6x = \underline{6x . a^{3x^2} \log_e a}$$

Expression (A) is now repeated

$$\frac{dy}{dx} = \frac{dy}{du} \times \frac{du}{dx} = y \log_e a \times \frac{du}{dx}$$

$$\therefore \frac{1}{y} \cdot \frac{dy}{dx} = \log_e a \times \frac{du}{dx}$$

That is, $\frac{1}{y} \cdot \frac{dy}{dx} = \log_e a$ multiplied by the derivative of $u$ (the function of $x$ which constitutes the exponent of $a$).

The use of this equation makes it possible to calculate the derivative of the function $a^n$ (where $n$ is a function of $x$) without having to work through all the substitutions.

Thus for the function

$$y = a^{5x^3}$$

$$\frac{1}{y} \cdot \frac{dy}{dx} = \log_e a \times 15x^2$$

$$\therefore \frac{dy}{dx} = y \log_e a \times 15x^2$$

$$= a^{5x^3} \log_e a \times 15x^2$$

This principle of logarithmic differentiation may be extended to simplify the differentiation of products, quotients, roots, and other complicated functions. As an example let

$$\log_e y = 3x^2 = u$$

Then $\quad u = 3x^2; \quad$ whence $\frac{du}{dx} = 6x$

and $\quad u = \log_e y; \quad$ whence $\frac{du}{dy} = \frac{1}{y}$

or $\quad \frac{dy}{du} = y$

but $\quad \frac{dy}{dx} = \frac{du}{dx} \times \frac{dy}{du} = \frac{du}{dx} \times y$

$$\therefore \frac{1}{y} \cdot \frac{dy}{dx} = \frac{du}{dx} = 6x$$

$$\therefore \frac{dy}{dx} = 6xy$$

This result shows that if we have a function of the form $\log_e y = u$ (a function of $x$), then

$$\frac{1}{y}\cdot\frac{dy}{dx} = \frac{du}{dx}$$

and the desired derivative $\frac{dy}{dx}$ may be obtained by multiplying $\frac{du}{dx}$ by $y$. To use this method we first express a function in terms of logarithms: If

$$y = (x^2-5)(x^3+7)$$

let $\quad u = (x^2-5); \quad$ whence $\frac{du}{dx} = 2x$

and $\quad v = (x^3+7); \quad$ whence $\frac{dv}{dx} = 3x^2$

then $\quad y = uv$

The logarithm of one side of the equation is equal to the logarithm of the other side. That is

$$\log_e y = \log_e u + \log_e v$$

This is in the form required as the right-hand side of the equation is a function of $x$. Then

$$\frac{1}{y}\cdot\frac{dy}{dx} = \text{derivative of } \log_e u + \text{derivative of } \log_e v \qquad \text{(i)}$$

*Derivative of $f(u) = \log_e u$*

The derivative of $\quad \log_e u = \frac{1}{u} \quad$ with respect to $u$

$$= \frac{1}{u}\cdot\frac{du}{dx} \quad \text{,, \quad ,, \quad ,, } x$$

$$= \frac{1}{x^2-5}2x = \frac{2x}{x^2-5}$$

*Derivative of $f(v) = \log_e v$*

The derivative of $\log_e v = \dfrac{1}{v}$ with respect to $v$

$$= \frac{1}{v}\cdot\frac{dv}{dx} \quad ,, \quad ,, \quad ,, \; x$$

$$= \frac{1}{x^3+7}3x^2 = \frac{3x^2}{x^3+7}$$

So equation (i) becomes

$$\frac{1}{y}\cdot\frac{dy}{dx} = \frac{2x}{x^2-5} + \frac{3x^2}{x^3+7}$$

and
$$\frac{dy}{dx} = y\left(\frac{2x}{x^2-5} + \frac{3x^2}{x^3+7}\right)$$

$$= (x^2-5)(x^3+7)\left(\frac{2x}{x^2-5} + \frac{3x^2}{x^3+7}\right)$$

$$= 5x^4 - 15x^2 + 14x$$

This method would not be required in this particular example because it would be easier to use either the product rule or to multiply out the expression $(x^2-5)(x^3+7) = x^5-5x^3+7x^2-35$ and to differentiate this by first principles

$$\frac{dy}{dx} = 5x^4 - 15x^2 + 14x$$

This procedure, once it is understood, may be employed in more difficult calculations where the simpler methods would be inadequate.

## *Chapter 12*

# INTEGRATION METHODS

The elements of the integral calculus were demonstrated in Chapter 9 and a method was shown for integrating a simple expression. Simple expressions however are relatively few in number and it is necessary to consider the integration of more complicated expressions.

It has already been noted that the process of integration cannot be reduced to a set of general rules. A number of basic rules may be followed for certain expressions, but a difficulty arises in that these rules cannot be applied generally to each and every expression. The art of integration lies in the ability to recognize the different types of expression.

Some expressions may be integrated in one way, others in another way, some cannot be integrated at all. This does not necessarily mean that a specific integrand has no related integral; it may merely mean that we cannot find out what the latter is. Yet, even if an integrand really has no integral, this in itself is not a self-contradiction of the calculus although it may indicate its limitations.

A similar kind of limitation exists in ordinary arithmetic. Every whole number may be squared, and the square number obtained will also be a whole number. But it does not follow that, reversing the process, the square root of every whole number is also a whole number. The square root of 10, for example, is not a whole number.

Our approach to integration must be adaptable and flexible. The actual method we adopt must depend upon the nature of the integrand. Some few integrands may be integrated simply by inspection. Since integration is the inverse of differentiation we may be able to recognize the function from which the integrand is derived. For example if we have to integrate $\int 2x\,dx$, we may associate this at once with the derivative

$$\frac{dy}{dx} = 2x$$

and we can immediately recognize that this is the derivative of $y = x^2+c$. Accordingly

$$y = \int 2x\,dx = x^2+c$$

If the function $y$ cannot be recognized at once by inspection, it may still be possible to integrate an expression by reversing the mathematics of differentiation. Where this may prove prolonged, we may instead be able to use special formulae by reference to tables of integrals which have already been worked out for us. Yet another way is to pass the problem on to someone else! All the methods depend upon the basic concept of reversing the process of differentiation; the different methods merely join the process at different stages.

Before proceeding to consider the arithmetic of integration, it is desirable to remember that, unless we are integrating a function between limits, we are always searching for the indefinite integral and that this integral will always include the undetermined constant of integration (usually shown as $c$).

It has been shown that the function $y = ax^n$ has as its derivative $\frac{dy}{dx} = nax^{n-1}$; also that, if we have an integrand of this form, we can integrate it by using the reverse process

$$y = \int nax^{n-1}\,dx = ax^n+c$$

When the function is differentiated the constant $a$ is carried through to the derivative. Consequently when we integrate $\int nax^{n-1}\,dx$, or $\int a(nx^{n-1})\,dx$, it will be apparent that the constant $a$ must also appear in the integral.

We now repeat that $\int a(nx^{n-1})\,dx$ may be interpreted as 'find the limiting value of the sum of the quantities $a(nx^{n-1})\,dx$ for all values of $x$'. In doing this the hard way we would have to multiply every value like $(nx^{n-1})\,dx$ by $a$ and then add all these products. It is easier however to add all the values like $(nx^{n-1})\,dx$ and then to multiply their total by $a$. The same result is obtained, and this operation is represented as

$$y = a \int (nx^{n-1})\, dx$$

We have merely transferred the constant $a$ to the left-hand side of the integral sign. Thus, we need only integrate $\int (nx^{n-1})\, dx$ as a first stage and then, finally, multiply by $a$. For example

$$\int 10x^4\, dx = 10 \int x^4\, dx$$

The integrand $\int x^4\, dx$ may be integrated by the method shown in Chapter 9. Add 1 to the exponent of $x$; and divide the expression by a number equivalent to the number of the new exponent. Then

$$\int x^4\, dx = \frac{x^5}{5} + c_1$$

and $$\int 10x^4\, dx = \frac{10x^5}{5} + 10c_1 = 2x^5 + c_2$$

In doing this we transferred the constant factor 10 to the left side of the integral sign. Only constants may be transferred in this way. No term involving $x$ can be transferred since this would give a meaningless result. Thus the integrand $\int (x^3+2x^2)\, dx$ may be expressed instead as $\int x^2(x+2)\, dx$. It cannot however be expressed as $x^2 \int (x+2)\, dx$ because $x^2$ is not constant. If we tried to do this we would first integrate $\int (x+2)\, dx$ and would then have to multiply the result by the numerical equivalent of $x^2$. But this is meaningless since $x$ is a variable quantity and accordingly $x^2$ cannot have one constant numerical equivalent.

In the above demonstrations, in which integration is treated as the inverse of differentiation, we have used a notation which suggests that differentiation is the starting point of all the calculations. Thus in the function $y = x^n$ the exponent of $x$ is represented by $n$. The derivative is $nx^{n-1}$ and the integration $\int nx^{n-1}\, dx$

brings us back to the original function. In the integrand however we start with an exponent of $x$ equal to $n-1$. Since the object of integration is to calculate the value of the original function we do not at first know the value of $n$ in the function, but we do know the value of the exponent of $x$ in the integrand and it is a more logical procedure to equate this exponent to $n$.

The symbol $n$ is merely part of a form of shorthand representing a number and has no other meaning whatever although the number which it represents may, as in the present instance, have a special significance. Provided we clearly define its use and impose relevant restrictions on its use in a particular problem, the symbol may be made to represent any number. In dealing with integration there is no point in introducing complications by beginning with a value of $n$ relevant to the original function which we as yet do not know. It is better to begin with a value of $n$ relevant to the integrand which we do know.

Thus we might have $\int x^n\,dx$ and integrating this would give

$$\int x^n\,dx = \left(\frac{1}{n+1}\right)x^{n+1}+c$$

Similarly

$$\int ax^n\,dx = \left(\frac{a}{n+1}\right)x^{n+1}+c$$

This represents the true inverse of differentiation quite clearly. It is easy to see from these expressions that to obtain the exponent of $x$ in the integral it is necessary to increase by 1 the exponent in the integrand, and that the coefficient of $x$ in the integrand must be divided by a number equal to the increased exponent in order to obtain the coefficient of $x$ in the integral.

## INTEGRATION OF A SUM

If the integrand is a sum of terms, then it may be fully integrated by integrating each term separately and then taking the sum of the separate integrals. This follows from the fact that the differentiation of the sum of two or more terms is achieved by taking the sum of the derivatives of the separate terms.

Thus if

$$y = \tfrac{1}{4}x^4+x^3$$

then

$$\frac{dy}{dx} = x^3+3x^2$$

so $$dy = x^3\,dx + 3x^2\,dx = (x^3+3x^2)\,dx$$

Then, on integrating, we simply reverse the whole process

$$\begin{aligned} y &= \int (x^3+3x^2)\,dx \\ &= \int x^3\,dx + \int 3x^2\,dx \\ &= (\tfrac{1}{4}x^4+c_1)+(x^3+c_2) \end{aligned}$$

It will be noted that in this integration two constants have appeared, since each separate integration must take into account the possibility of the existence of an added constant in the original function. These two constants may however in the final form of the indefinite integral be represented as one constant since the total of two constants is also constant. Then

$$\int (x^3+3x^2)\,dx = \tfrac{1}{4}x^4+x^3+c$$

Equally simple is the integration of an integrand expressed as the difference between terms. This is treated in exactly the same way except that there is a negative sign.

Thus

$$\begin{aligned} \int (x^3-3x^2)\,dx &= (\tfrac{1}{4}x^4+c_1)-(x^3+c_2) \\ &= \tfrac{1}{4}x^4-x^3+c \end{aligned}$$

In this instance the final constant $c$ is of course the difference between the two constants $c_1$ and $c_2$ and may be negative or positive.

An integrand may itself consist of one or more terms involving $x$ and one constant term. Thus, where $k$ is constant, let

$$\begin{aligned} y &= \int (x^3+k)\,dx \\ &= \int x^3\,dx + \int k\,dx \end{aligned}$$

The integral of $\int x^3\,dx$ is, as shown above, $\tfrac{1}{4}x^4+c$. The integral of $\int k\,dx$ may readily be calculated if it is remembered that $k$ is the derivative of the function $y = kx$. That is, $\frac{dy}{dx} = k$, or $dy = k\,dx$. Reversing the process, if $dy = k\,dx$, then $y = kx+c$.

Accordingly the total integral of the above expression

$$\int x^3\,dx + \int k\,dx = (\tfrac{1}{4}x^4+c_1)+(kx+c_2)$$
$$= \tfrac{1}{4}x^4+kx+c$$

## INTEGRATION OF A PRODUCT

A further case to consider is that of an integrand in the form of a product, such as $\int (x+2)(x+4)\,dx$. We have not seen any derivative in the form of a product and accordingly we cannot integrate directly. This difficulty may be resolved by first multiplying out so that the integrand is expressed as a sum of terms. Thus

$$\begin{aligned} y &= \int (x+2)(x+4)\,dx \\ &= \int (x^2+6x+8)\,dx \\ &= \int x^2\,dx + \int 6x\,dx + \int 8\,dx \\ &= \tfrac{1}{3}x^3+3x^2+8x+c \end{aligned}$$

## INTEGRATION OF A QUOTIENT

Consider the following:

$$y = \int \left(\frac{x^5+4x^3+2x^2}{x^3}\right)dx$$

In this instance it is first necessary to split the fraction into its component parts. These may then be integrated in the usual way

$$\begin{aligned} y &= \int \frac{x^5}{x^3}dx + \int \frac{4x^3}{x^3}dx + \int \frac{2x^2}{x^3}dx \\ &= \int x^2\,dx + \int 4\,dx + 2\int \frac{1}{x}dx \\ &= \tfrac{1}{3}x^3+4x+2\log_e x+c \end{aligned}$$

The principle employed above may be extended to more complicated integrands. The denominator in the above example is $x^3$

and it is possible to simplify the component fractions. If however we have an integrand of the form

$$\int\left(\frac{x^2}{x-1}\right)dx$$

there is the problem that we cannot divide $x^2$ by $(x-1)$ to give a simple method of working, but we can transform the numerator by the expedient of deducting 1 and then adding it back again. Integration then becomes comparatively easy.

$$y = \int\left(\frac{x^2}{x-1}\right)dx = \int\left(\frac{(x^2-1)+1}{x-1}\right)dx$$

$$= \int\left(\frac{x^2-1}{x-1}+\frac{1}{x-1}\right)dx$$

$$= \int\left[(x+1)+\frac{1}{x-1}\right]dx$$

$$= \int(x+1)dx+\int\left(\frac{1}{x-1}\right)dx$$

$$= \tfrac{1}{2}x^2+x+\log_e(x-1)+c$$

## OTHER INTEGRANDS

In reversing the process of differentiation it is of course always necessary to be sure that we are following the right process. We may, for example, have to integrate $y = \int x^{-1}\,dx$. The original function $y$ is such that its derivative is $x^{-1}$. It has already been shown* that no derivative of this form can be obtained by the ordinary method of differentiating a simple expression involving $x$. It was subsequently seen† however that the derivative $x^{-1}$ was obtained by differentiating the function

$$y = \log_e x$$

Consequently, by reversing the process, the integration of $\int x^{-1}\,dx$ will give the value $y = \log_e x+c$.

Similarly if $y = \int ax^{-1}\,dx$ then integration will give the value $y = a\log_e x+c$.

* Page 54. † Page 146.

These results are possible because we are aware of the process of differentiation and may therefore reverse it. If we had not known that differentiating the function $y = \log_e x$ gives the derivative $x^{-1}$ we would not have been able to integrate $\int x^{-1}\,dx$ at all. There are many expressions which cannot be identified as derivatives of specific functions and accordingly if such expressions appear as part of an integrand then we shall be unable to integrate the latter.

## SUCCESSIVE INTEGRATION

Just as differentiation may be continued successively until eventually a constant derivative is obtained, so may integration be continued successively. Thus if $\frac{d^2y}{dx^2} = x^4$ then by two integrations we may express $y$ in terms of $x$. We originally obtain $\frac{d^2y}{dx^2}$ by differentiating $\frac{dy}{dx}$. So when we integrate $\int \left(\frac{d^2y}{dx^2}\right) dx$ we obtain the first derivative $\frac{dy}{dx}$. This in turn may be integrated to give the value of $y$.

Let $$\frac{d^2y}{dx^2} = x^4$$

then $$\frac{dy}{dx} = \int x^4\,dx$$
$$= \tfrac{1}{5}x^5 + c_1$$

then $$y = \int (\tfrac{1}{5}x^5 + c_1)dx$$
$$= \int \tfrac{1}{5}x^5\,dx + \int c_1\,dx$$
$$= \frac{x^6}{30} + c_1x + c_2$$

Here we have had to introduce two constants $c_1$ and $c_2$. The constant $c_1$ arises in the first integration. The constant $c_2$ arises in the second integration. If we wish to find the complete numerical relationship between $x$ and $y$ we shall need to have two pairs of corresponding values of $x$ and $y$. These will permit us to set up two simultaneous equations from which the values of $c_1$ and $c_2$ may be obtained.

If, for example, two such values are

(i) $x = 3; \quad y = 49 \cdot 3$

(ii) $x = 6; \quad y = 1595 \cdot 2$

Then, from (i)

$$\frac{3^6}{30} + 3c_1 + c_2 = 49 \cdot 3$$

whence $$3c_1 + c_2 = 25 \cdot 0 \qquad \text{(iii)}$$

and from (ii) $$\frac{6^6}{30} + 6c_1 + c_2 = 1595 \cdot 2$$

whence $$6c_1 + c_2 = 40 \cdot 0 \qquad \text{(iv)}$$

and from (iii) and (iv)

$$6c_1 + c_2 = 40$$
$$3c_1 + c_2 = 25$$

Subtracting: $$3c_1 = 15$$

whence $$c_1 = 5$$

and $$c_2 = 10$$

The true function of $x$ given by integration in this instance is therefore

$$y = \frac{x^6}{30} + 5x + 10$$

## PARTIAL FRACTIONS

In Chapter 10 it was seen that the differentiation of a fraction could very often be achieved most easily by first splitting it into partial fractions. Much the same device may be employed in integrating.

For example, the fraction $\frac{3x+1}{x^2-1}$ may be expressed differently as (see Appendix I):

$$\frac{3x+1}{x^2-1} = \frac{1}{x+1} + \frac{2}{x-1}$$

$$\therefore \int\left(\frac{3x+1}{x^2-1}\right)dx = \int\left(\frac{1}{x+1}\right)dx + \int\left(\frac{2}{x-1}\right)dx$$

$$= \log_e(x+1) + 2\log_e(x-1) + c$$

In this instance we have a denominator $x^2-1$ which is the difference between two squares. The method is not confined to fractions of this form. Provided the numerator is itself linear (i.e. contains no exponent of $x$ greater than 1) and provided also that the denominator may be factorized into linear factors then no difficulties will normally arise. Thus

$$\int\left(\frac{8x+14}{x^2+2x-8}\right)dx = \int\left(\frac{5}{x-2}\right)dx + \int\left(\frac{3}{x+4}\right)dx$$

$$= 5\log_e(x-2) + 3\log_e(x+4) + c$$

If the numerator is not linear, the normal procedure is to divide it first by the denominator and to use the method of partial fractions only in respect of the remainder. Thus

$$\int\left(\frac{x^2+10x+6}{x^2+2x-8}\right)dx = \int\left(1+\frac{8x+14}{x^2+2x-8}\right)dx$$

$$= \int 1\,dx + \int\left(\frac{8x+14}{x^2+2x-8}\right)dx$$

$$= \int 1\,dx + \int\left(\frac{5}{x-2}\right)dx + \int\left(\frac{3}{x+4}\right)dx$$

$$= x + 5\log_e(x-2) + 3\log_e(x+4) + c$$

There are many ingenious methods for dealing with the integration of more complicated expressions. The foregoing will be sufficient to show that some ingenuity is certainly required and that it pays off handsomely.

Two other useful methods are integration by substitution and integration by parts.

### SUBSTITUTION

The device of substitution employed in differentiation may also be employed in integration. It is, for example, most useful in the integration of square roots.

$$y = \int \sqrt{2+x}\,dx$$

Let $\quad u = 2+x \quad$ whence $\dfrac{du}{dx} = 1 \quad$ (so $dx = du$)

then $$y = \int u^{\frac{1}{2}}\,du$$
$$= \tfrac{2}{3}u^{\frac{3}{2}}+c$$

But $\quad u = 2+x, \quad$ so

$$y = \tfrac{2}{3}(2+x)^{\frac{3}{2}}+c = \tfrac{2}{3}\sqrt{(2+x)^3}+c$$

Other forms of substitution involve the use of trigonometric or hyperbolic functions and are referred to later.

## INTEGRATION BY PARTS

Substitution is also employed in the process of integration by parts. This method is extremely useful in integrating products of two or more functions. The formula for the differentiation of the product $y = uv$ (where $u$ and $v$ are functions of $x$) is

$$\frac{dy}{dx} = u\frac{dv}{dx}+v\frac{du}{dx}$$

Reversing the process, then

$$(y) = uv = \int \left(u\frac{dv}{dx}\right)dx + \int \left(v\frac{du}{dx}\right)dx$$

so $$uv = \int u\,dv + \int v\,du$$

or $$\int u\,dv = uv - \int v\,du$$

If therefore we have an integrand which may be expressed in the form $u\,dv$ then we may equate $\int u\,dv$ to $uv - \int v\,du$. This will obviously be most useful when $\int v\,du$ is easier to integrate than is $\int u\,dv$, and in fact this is often found to be so. As an example, let

$$y = \int x^2 \log_e x\,dx$$

It will first be noted that this integrand may be written as either

$$x^2 \log_e x \, dx \qquad \text{(A)}$$

or

$$\log_e x . x^2 \, dx \qquad \text{(B)}$$

In expression B we have merely changed the order of the terms and this makes no difference to the value of the expression.

Either of these integrands may be transformed to $u \, dv$ where

*in integrand A:* $u = x^2$; $dv = \log_e x \, dx$

*in integrand B:* $u = \log_e x$; $dv = x^2 \, dx$

For a reason which will be explained later, we take integrand B for our present purpose

(i) $$u = \log_e x$$

Then by differentiating,

$$\frac{du}{dx} = \frac{1}{x}$$

and

$$du = \frac{1}{x} dx \qquad (\text{i}a)$$

(ii) $$dv = x^2 \, dx$$

Then by integrating,

$$v = \tfrac{1}{3}x^3 = \frac{x^3}{3} \qquad (\text{ii}a)$$

(iii) From (i*a*) and (ii*a*)

$$uv = (\log_e x)\left(\frac{x^3}{3}\right) \qquad (\text{iii}a)$$

and

$$v \, du = \left(\frac{x^3}{3}\right)\left(\frac{1}{x} dx\right) = \frac{x^2}{3} dx \qquad (\text{iii}b)$$

Now

$$\int u \, dv = uv - \int v \, du$$

$$= \frac{x^3}{3} \log_e x - \int \frac{x^2}{3} dx$$

$$= \frac{x^3}{3} \log_e x - \frac{x^3}{9} + c$$

or

$$\frac{x^3}{3}(\log_e x - \tfrac{1}{3}) + c$$

and this is the desired indefinite integral of $\int x^2 \log_e x\,dx$ since we originally set this as being equal to $\int u\,dv$.

The reason for choosing the integrand form B is simple. This choice makes the calculation an easier one! Instead of integrating $\int x^2 \log_e x\,dx$ direct, we have been able to transform the required calculation so that it involves only the integration of the much simpler $\int \frac{x^2}{3} dx$.

If we had chosen integrand form A, then we would have had

$$dv = \log_e x\,dx$$

whence
$$v = \int \log_e x\,dx = x(\log_e x - 1) \qquad \text{(D)}$$

(See end of chapter)

also
$$u = x^2$$

whence
$$du = 2x\,dx$$

so
$$\int v\,du = \int x(\log_e x - 1)2x\,dx$$
$$= \int 2x^2(\log_e x - 1)\,dx$$

This is even more complicated than the original integrand $\int x^2 \log_e x\,dx$ and we would have gained nothing by our substitution.

A great deal thus depends upon which function is chosen to be represented by $u$. So far as it is possible it is obviously desirable to be able to look ahead a little when making the choice to judge which selection will produce the easier integral. It will however be noted that $u$ is the function which first has to be differentiated. In many cases if we choose the function which is easier to differentiate and represent this function as $u$, then the final integration required will also be easier. This however cannot be said to be a general rule and the ability to make the right choice depends largely upon our experience.

It was noted in equation D above that

$$\int \log_e x\, dx = x(\log_e x - 1) + c$$

This has not yet been demonstrated and, as it is a very important result, the proof is now given. We again use the method of integration by parts.

Let $$\int \log_e x\, dx = \int u\, dv$$

then $$u = \log_e x \qquad \text{whence } du = \frac{1}{x}dx$$

and $$dv = dx \qquad \text{whence } v = x$$

Substituting the various equivalents in the equation

$$\int u\, dv = uv - \int v\, du$$

$$\int \log_e x\, dx = x \log_e x - \int (x)\left(\frac{1}{x}\right)dx$$

$$= x \log_e x - \int dx$$

$$= x \log_e x - x + c$$

$$= x(\log_e x - 1) + c$$

*Chapter 13*

# TRIGONOMETRIC FUNCTIONS

In many applications of the calculus we have to deal with functions which are expressed in trigonometrical terms. The study of trigonometric functions does not introduce any new calculus ideas but it does involve a certain amount of use of transformations of the functions themselves in order to make the process of differentiation easier. For those whose trigonometry has become a little rusty, the basic ratios and equivalents are included in Appendix II.

In the equations

$$y = \cos\theta$$

and $$y = \sin\theta$$

the variable $y$ is expressed as functions of the independent variable $\theta$ which represents the measurement of an angle. As $\theta$ varies so $y$ also varies and the functions may be differentiated so as to reveal the rate of change of $y$ relative to the change of $\theta$.

We could just as well have used the symbol $x$ instead of $\theta$, but it is traditional in mathematics that angles are usually denoted by Greek letters and accordingly our examples employ the letter $\theta$ (or 'theta') to represent a variable angle.

It is instructive to draw the curves of these two functions. (See Fig. 34.) At once we can see a remarkable similarity between the curves. Their general pattern is identical but they occupy different positions relative to the axes. The sections of the curves illustrated repeat indefinitely and the functions $\sin\theta$ and $\cos\theta$ are called periodic functions.

The measurements of the angle $\theta$ are shown in degrees along the axis. In the calculus however angles are measured in *radians*. This form of measurement is referred to as the circular measure of an angle because it relates the measurement to the properties of a circle. A radian is the angle subtended at the centre of a circle by an arc which is equal in length to the radius of the circle. In Fig. 35 OA is a radius of the circle. The length of the arc AB is exactly equal to OA. The angle $\theta$ then measures 1 radian.

The ratio of the circumference of a circle to its radius is such that the radius may be marked off the circumference $2\pi$ times. The

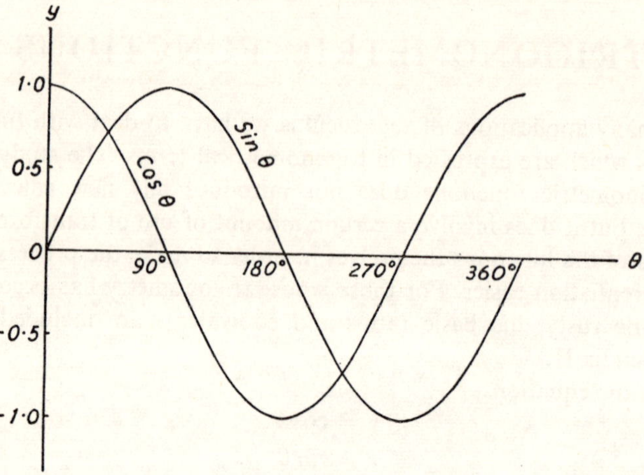


FIG. 34

circumference may therefore be regarded as consisting of $2\pi$ arcs each of which is equal in length to the radius. Each arc subtends an angle of 1 radian. Consequently $2\pi$ arcs together subtend a

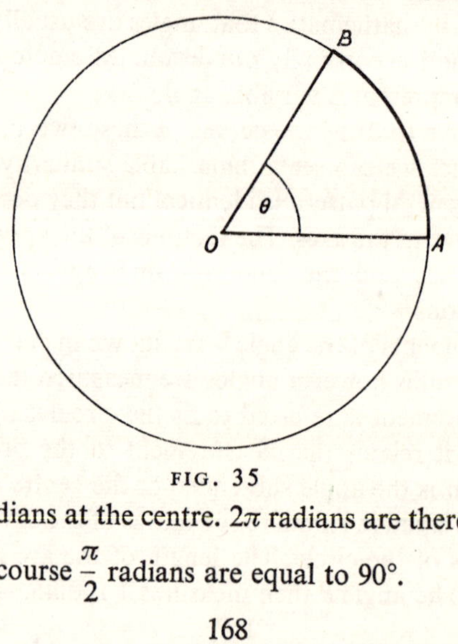


FIG. 35

total of $2\pi$ radians at the centre. $2\pi$ radians are therefore equal to 360°; and of course $\frac{\pi}{2}$ radians are equal to 90°.

The degree method of measurement is useful in geometry where the measurement of the angle is itself the primary object of calculations. In the calculus however, as also in trigonometry, the size of an angle may be only one of many inter-related dimensions. The usefulness of the radian as a unit of measurement may at once be appreciated. By definition OA = AB. Consequently $\frac{BA}{OA} = 1$, and it is this ratio which measures the angle $\theta$ in radians. As the length of the arc is increased so also is the angle $\theta$ increased, but the measurement of $\theta$ in radians will always be given by the ratio

$$\theta = \frac{\text{arc}}{\text{radius}} \qquad \text{or} \qquad \text{arc} = \theta \times \text{radius}$$

and this of course applies to any circle.

If then we know the radius of a circle and can measure the angle $\theta$ we can also use this simple ratio to calculate the length of the arc. Such a calculation would be unnecessarily complicated if we were to introduce arbitrary methods of measurement. We are accustomed to dealing with right angles and the system whereby they are divided into ninety degrees and then again into minutes and seconds. But this division is purely arbitrary. A right angle is divided into ninety degrees only because it was considered at the time when the division was first established that it was convenient to divide the angle into ninety parts.

This system is not convenient for calculus methods and we must discard it in favour of circular measure which was designed to be convenient. In doing this we are changing only the method of measuring the angle, just as we may measure a distance in miles or in kilometres. The angle itself is not affected in any way. The radian is clearly defined in terms of the other dimensions of a circle and this definite relationship enables us to transform a number of awkward problems into forms which we can handle more conveniently.

Before proceeding to consider the differentiation of trigonometric functions, it is first necessary to determine the limiting value of the ratio $\frac{\sin\theta}{\theta}$. The reason for this will become apparent later.

This ratio is between the sine of the angle $\theta$ and the angle itself measured in radians.

In Fig. 36, the point O is at the centre of a circle the radius of which is unity. ABC is an arc of the circle and AC is the chord of this arc. The radius OB is drawn at right angles to AC, so as to bisect the latter at D, and is then extended to E to meet the

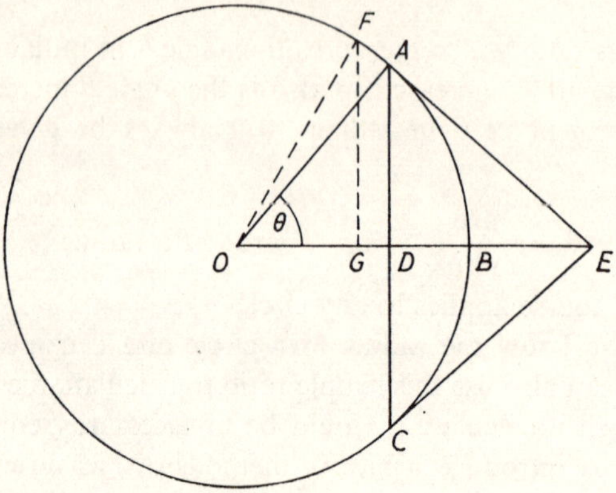


FIG. 36

tangents drawn at points A and C respectively. The angle BOA $= \theta$ radians. Inspection of the construction shows that

$$\text{EA}+\text{EC} > \text{arc ABC}$$

$$\text{arc ABC} > \text{chord AC}$$

But EA $=$ EC. Then

$$2\text{EA} > \text{arc ABC} > \text{chord AC}$$

and

$$\text{EA} > \text{arc AB} \quad > \text{AD}$$

We now find three equivalents as follows, remembering that OA is a radius of the circle and that, therefore, OA $= 1$.

(i) By circular measure the angle $\theta$, in radians,

$$= \frac{\text{arc AB}}{\text{OA}} = \text{arc AB}$$

(ii) $$\sin\theta = \frac{\text{AD}}{\text{OA}} = \text{AD}$$

(iii) $$\tan\theta = \frac{\text{EA}}{\text{OA}} = \text{EA}$$

Accordingly the inequalities

$$EA > \text{arc } AB > AD$$

become $$\tan\theta > \quad \theta \quad > \sin\theta$$

But $$\tan\theta = \frac{\sin\theta}{\cos\theta} \qquad \text{(see Appendix II)}$$

Therefore $$\frac{\sin\theta}{\cos\theta} > \theta > \sin\theta$$

and dividing by $\sin\theta$: $$\frac{1}{\cos\theta} > \frac{\theta}{\sin\theta} > 1$$

But when $\theta$ approaches zero, $\cos\theta$ approaches the value 1 and therefore $\frac{1}{\cos\theta}$ also approaches 1. In the limit therefore the ratio $\frac{\theta}{\sin\theta}$ cannot be greater nor less than 1. In other words (since the reciprocal of 1 is also 1):

$$\lim_{\theta\to 0} \frac{\sin\theta}{\theta} = 1$$

It will have been noted that in order to achieve this result a fair amount of substituting, transposing, and transforming of expressions is involved. Herein lies the art of the calculus as distinct from its strictly scientific bases.

## SINES

The result just achieved is extremely important and clears the way to an investigation of the function

$$y = \sin\theta$$

Reference is again made to Fig. 36. The sine of the angle $\theta$ is equivalent to the ratio $\frac{AD}{OA}$. Hence $y = \frac{AD}{OA}$. If again we assume that the radius OA is unity, then $y = AD$. We now increase the size of the angle by adding to it the indefinitely small increment $\Delta\theta$ (this being the angle FOA). The sine of the angle $\theta+\Delta\theta$ is now $\frac{FG}{OF}$. OF is a radius and is therefore equal to unity. Hence the sine of $\theta+\Delta\theta$ = FG. This is greater than AD (= $y$ when the angle is $\theta$) and, in fact, represents $y$ when the angle is $\theta+\Delta\theta$. The value of $y$ has increased because of the increase in the size of the angle. This increment in $y$ is again represented by $\Delta y$ and the following equation represents the changed circumstances

$$y+\Delta y = \sin(\theta+\Delta\theta)$$

but $$y = \sin\theta$$

therefore $$\Delta y = \sin(\theta+\Delta\theta)-\sin\theta$$

This represents $\Delta y$ as the difference between two sines. We here use a trigonometrical equivalent (see Appendix II) which states that, for any two angles $M$ and $P$, the difference between their sines may otherwise be represented as

$$\sin M-\sin P = 2\cos\left(\frac{M+P}{2}\right).\sin\left(\frac{M-P}{2}\right)$$

If we substitute $\theta+\Delta\theta$ and $\theta$ for $M$ and $P$ respectively, we have

$$\Delta y = 2\cos\left(\frac{\theta+\Delta\theta+\theta}{2}\right)\sin\left(\frac{\theta+\Delta\theta-\theta}{2}\right)$$

$$= 2\cos\left(\theta+\frac{\Delta\theta}{2}\right).\sin\left(\frac{\Delta\theta}{2}\right)$$

$$\therefore \frac{\Delta y}{\Delta\theta} = \frac{2\cos\left(\theta+\frac{\Delta\theta}{2}\right).\sin\frac{\Delta\theta}{2}}{\Delta\theta}$$

$$= \cos\left(\theta+\frac{\Delta\theta}{2}\right)\times\frac{\sin\frac{\Delta\theta}{2}}{\frac{\Delta\theta}{2}}$$

The second factor in this expression is in the form of the ratio of the sine of an angle to the size of the angle itself. This is the ratio which was investigated at the beginning of this chapter, where it was noted that the limiting value of the ratio is equal to 1 as the angle approaches zero. In the present ratio the angle is $\frac{\Delta\theta}{2}$ and accordingly the limiting value of the ratio is equal to 1 as $\frac{\Delta\theta}{2}$ approaches zero.

Furthermore as $\frac{\Delta\theta}{2}$ approaches zero, so $\theta+\frac{\Delta\theta}{2}$ approaches $\theta$ and, consequently, $\cos\left(\theta+\frac{\Delta\theta}{2}\right)$ approaches $\cos\theta$. For the limiting value therefore

$$\frac{\Delta y}{\Delta\theta} = \cos\left(\theta+\frac{\Delta\theta}{2}\right)\times\frac{\sin\frac{\Delta\theta}{2}}{\frac{\Delta\theta}{2}}$$

becomes $$\frac{dy}{d\theta} = \cos\theta\times 1 = \cos\theta$$

So the derivative of a function $y$ expressed as the sine of a variable angle is equal to the cosine of that angle.

Thus $$y = \sin\theta$$

$$\frac{dy}{d\theta} = \cos\theta$$

This is indeed a remarkable result and in it we may recognize one of the astonishing inter-related phenomena of mathematics which appear to have a magic of their own.

## COSINES

The nature of this relation between the cosine and the derivative of the sine having been revealed, it will probably come as less of a surprise to learn that the derivative of the cosine may be as elegantly expressed in terms of the sine.

Using a similar method as before

$$y = \cos\theta$$

$$\therefore\ y+\Delta y = \cos(\theta+\Delta\theta)$$

$$\therefore\ \Delta y = \cos(\theta+\Delta\theta)-\cos\theta$$

$\Delta y$ is now expressed as the difference between two cosines. Here again we use a trigonometrical equivalent (see Appendix II)

$$\cos M-\cos P = -2\sin\left(\frac{M+P}{2}\right)\sin\left(\frac{M-P}{2}\right).$$

If we substitute $\theta+\Delta\theta$ and $\theta$ for $M$ and $P$ respectively then

$$\Delta y = -2 \sin\left(\frac{\theta+\Delta\theta+\theta}{2}\right) \sin\left(\frac{\theta+\Delta\theta-\theta}{2}\right)$$

$$= -2 \sin\left(\theta+\frac{\Delta\theta}{2}\right) \sin\frac{\Delta\theta}{2}$$

Divide both sides by $\Delta\theta$, then

$$\frac{\Delta y}{\Delta\theta} = -2 \sin\left(\theta+\frac{\Delta\theta}{2}\right) \times \frac{\sin\frac{\Delta\theta}{2}}{\Delta\theta}$$

$$= -\sin\left(\theta+\frac{\Delta\theta}{2}\right) \times \frac{\sin\frac{\Delta\theta}{2}}{\frac{\Delta\theta}{2}}$$

The second factor is once more familiar and we know that its limiting value, as $\frac{\Delta\theta}{2}$ approaches zero, is 1. And as $\frac{\Delta\theta}{2}$ approaches zero, so $\left(\theta+\frac{\Delta\theta}{2}\right)$ approaches $\theta$ and consequently $\sin\left(\theta+\frac{\Delta\theta}{2}\right)$ approaches $\sin\theta$. Thus the limiting value:

$$\frac{dy}{d\theta} = -\sin\theta \times 1 = -\sin\theta$$

That is, the derivative of the cosine of an angle is equal to the sine of the angle but with the sign changed from positive to negative.

## TANGENTS

To find the derivative of $y = \tan\theta$, we could again proceed from first principles, but it is easier to derive the value from the known relationships between the two trigonometrical functions already studied. The tangent of an angle is related to the sine and cosine as

$$\tan\theta = \frac{\sin\theta}{\cos\theta}$$

To differentiate tan $\theta$ therefore we may use the quotient rule

$$\frac{dy}{d\theta}=\frac{v\dfrac{du}{d\theta}-u\dfrac{dv}{d\theta}}{v^2}$$

where $u = \sin\theta$, and $v = \cos\theta$. From the foregoing demonstration we know that $\frac{du}{d\theta}$, the derivative of $\sin\theta$, $= \cos\theta$; and that $\frac{dv}{d\theta}$, the derivative of $\cos\theta$, $= -\sin\theta$. Consequently the expression for $\frac{dy}{d\theta}$ derived from the quotient rule becomes:

$$\frac{dy}{d\theta}=\frac{(\cos\theta)(\cos\theta)-(\sin\theta)(-\sin\theta)}{(\cos\theta)^2}.$$

$$=\frac{\cos^2\theta+\sin^2\theta}{\cos^2\theta}$$

But $\cos^2\theta+\sin^2\theta = 1$ (see Appendix II)

therefore $\dfrac{dy}{d\theta}=\dfrac{1}{\cos^2\theta}$ or $\sec^2\theta$

## OTHER FUNCTIONS

The derivatives of the secant, the cosecant, and the cotangent (i.e. the reciprocals of the cosine, sine, and tangent respectively) may all be derived by the use of the quotient rule. It is not proposed to give the working here but the derivatives are summarized for convenience.

| *Function* | *First derivative* |
|---|---|
| $\sin\theta$ | $\cos\theta$ |
| $\cos\theta$ | $-\sin\theta$ |
| $\tan\theta$ | $\sec^2\theta$ |
| $\operatorname{cosec}\theta$ | $-\operatorname{cosec}\theta.\cot\theta$ |
| $\sec\theta$ | $\sec\theta.\tan\theta$ |
| $\cot\theta$ | $-\operatorname{cosec}^2\theta$ |

This tabulation shows the derivative $\frac{dy}{d\theta}$ for the simple trigonometric functions. Fortunately there is no need to go through all the working every time. All that is necessary is that these results should be memorized so that they may be used as the occasion arises. In practice of course one will often encounter functions which are not of this simple form – that is, when they are in fact functions of a function. But just as, in the differentiation of ordinary algebraic functions, we resorted to the use of substitution so also is this possible in exactly the same way with the trigonometric functions. The method involved is the same; we are merely dealing with a different kind of function. A few examples follow.

*The derivative of* $y = \cos^3 \theta$

This is the same as $(\cos \theta)^3$

Let $\cos \theta = u$ whence $\frac{du}{d\theta} = -\sin \theta$

then $y = u^3$ whence by simple differentiation

$$\frac{dy}{du} = 3u^2 = 3 \cos^2 \theta$$

$$\frac{dy}{d\theta} = \frac{dy}{du} \times \frac{du}{d\theta} = (3 \cos^2 \theta)(-\sin \theta)$$

$$= \underline{-3 \cos^2 \theta . \sin \theta}$$

*The derivative of* $y = \frac{1}{4} \sin 2\theta$

Let $2\theta = u$ whence $\frac{du}{d\theta} = 2$

then $y = \frac{1}{4} \sin u$ whence $\frac{dy}{du} = \frac{1}{4} \cos u$

$$\frac{dy}{d\theta} = \frac{dy}{du} \times \frac{du}{d\theta} = \frac{1}{4} \cos u \times 2 = \underline{\frac{1}{2} \cos 2\theta}$$

*The derivative of* $y = 2 \sin \frac{\theta}{2}$

Let $\frac{\theta}{2} = u$ whence $\frac{du}{d\theta} = \frac{1}{2}$

then $y = 2 \sin u$ whence $\frac{dy}{du} = 2 \cos u$

$$\frac{dy}{d\theta} = \frac{dy}{du} \times \frac{du}{d\theta} = 2 \cos u \times \frac{1}{2} = \cos \frac{\theta}{2}$$

The successive derivatives of trigonometric functions are also very interesting. If $y = \sin\theta$, then $\frac{dy}{d\theta} = \cos\theta$. To obtain the second derivative of $\sin\theta$ we must differentiate $\frac{dy}{d\theta}$. That is, we differentiate $\cos\theta$ and the derivative of this has been shown to be $-\sin\theta$. To obtain the third derivative of $\sin\theta$, we must now differentiate $-\sin\theta$ and this derivative is $-\cos\theta$. Again, to obtain the fourth derivative, we differentiate $-\cos\theta$, and this derivative is $\sin\theta$. So we are back where we started! Thus we have:

$$y = \sin\theta$$

$$\frac{dy}{d\theta} = \cos\theta$$

$$\frac{d^2y}{d\theta^2} = -\sin\theta$$

$$\frac{d^3y}{d\theta^3} = -\cos\theta$$

$$\frac{d^4y}{d\theta^4} = \sin\theta$$

and these stages will repeat over and over again for successive derivatives. The same sort of ladder exists for the successive differentiation of $\cos\theta$ but this starts one rung lower.

There is another method whereby each of these successive derivatives may be expressed in sine form. For this we resort to another trigonometrical equivalent (see Appendix II):

$$\cos\theta = \sin\left(\theta+\frac{\pi}{2}\right)$$

Then if $\quad y = \sin\theta$

so $\quad \frac{dy}{d\theta} = \cos\theta = \underline{\sin\left(\theta+\frac{\pi}{2}\right)}$

$$\frac{d^2y}{d\theta^2} = \text{derivative of } \sin\left(\theta+\frac{\pi}{2}\right)$$

$$= \cos\left(\theta+\frac{\pi}{2}\right)$$

$$= \sin\left(\theta+\frac{\pi}{2}+\frac{\pi}{2}\right)$$

$$= \sin\left(\theta+\frac{2\pi}{2}\right)$$

This process may be repeated indefinitely with $\frac{\pi}{2}$ being added to the expression inside the brackets for each successive derivative and, generally, the $n$th derivative of $y = \sin\theta$ is

$$\frac{d^n y}{d\theta^n} = \sin\left(\theta+\frac{n\pi}{2}\right)$$

It is possible to transform successive derivatives of $y = \cos\theta$ in much the same way from the equivalent (see Appendix II)

$$-\sin\theta = \cos\left(\theta+\frac{\pi}{2}\right)$$

Then if $\quad y = \cos\theta$

so $\quad \dfrac{dy}{d\theta} = -\sin\theta = \cos\left(\theta+\dfrac{\pi}{2}\right)$

and in general

$$\frac{d^n y}{d\theta^n} = \cos\left(\theta+\frac{n\pi}{2}\right)$$

## INVERSE FUNCTIONS

The relationship which is expressed in the function $y = \sin x$ may also be expressed in the form

$$x = \sin^{-1} y$$

where $x$ is expressed as a function of $y$, instead of $y$ being expressed as a function of $x$. The second form is the inverse function of the first and it includes a special trigonometric notation. The superscript $-1$ is not an index number but instead forms part of the symbol $\sin^{-1}$, the symbol for the inverse function of a sine. Similarly $\cos^{-1}$ is the symbol for the inverse function of a cosine.

The symbolism of

$$x = \sin^{-1} y$$

must be interpreted as stating '$x$ is the angle of which $y$ is the sine'. There are in fact many angles whose sine is $y$; here we restrict our attention to acute angles lying in the first quadrant of a circle. To differentiate this function in which $x$ is expressed as a function of $y$, we need to calculate the value of $\frac{dx}{dy}$. We proceed as follows

$$x = \sin^{-1} y$$

$$\therefore\ y = \sin x$$

$$\therefore\ \frac{dy}{dx} = \cos x$$

but
$$\frac{dx}{dy} = \frac{1}{\frac{dy}{dx}} = \frac{1}{\cos x}$$

We now draw on another trigonometric equivalent (see Appendix II)

$$\cos x = \sqrt{1-\sin^2 x}$$

$$\therefore\ \frac{1}{\cos x} = \frac{1}{\sqrt{1-\sin^2 x}}$$

but
$$y = \sin x$$

$$\therefore\ y^2 = \sin^2 x$$

$$\therefore\ \frac{1}{\cos x} = \frac{1}{\sqrt{1-y^2}} \quad \left(= \frac{dx}{dy}\right)$$

This process has been greatly simplified here in order to demonstrate the basic principle involved.

In substituting $\cos x = \sqrt{1-\sin^2 x}$ we have taken the positive and not the negative value of the root without justifying our choice. For a full discussion of this point, the reader is referred to *Introduction to the Infinitesimal Calculus* by G. W. Caunt (Oxford U.P.), p. 199.

It will be necessary later, when dealing with the areas of curved surfaces, to know something about the treatment of the general function

$$y = \sin^{-1}\left(\frac{x}{a}\right)$$

This is differentiated as follows, based on the previous demonstration

$$y = \sin^{-1}\left(\frac{x}{a}\right)$$

Let $\frac{x}{a} = v \quad \left(\text{whence } \frac{dv}{dx} = \frac{1}{a}\right)$

then $y = \sin^{-1} v$

or $v = \sin y \quad \text{whence } \frac{dv}{dy} = \cos y$

$$\text{and} \quad \frac{dy}{dv} = \frac{1}{\cos y}$$

$$= \frac{1}{\sqrt{1-\sin^2 y}}$$

But

$$v = \sin y$$

$$\therefore v^2 = \sin^2 y$$

$$\therefore \frac{1}{\sqrt{1-\sin^2 y}} = \frac{1}{\sqrt{1-v^2}}$$

Then

$$\frac{dy}{dx} = \frac{dv}{dx} \times \frac{dy}{dv}$$

$$= \frac{1}{a} \times \frac{1}{\sqrt{1-v^2}}$$

$$= \frac{1}{a} \times \frac{1}{\sqrt{1-\left(\frac{x}{a}\right)^2}}$$

$$= \frac{1}{\sqrt{a^2-x^2}}$$

The foregoing is an example of the relationship between trigonometrical ratios and algebraic expressions. Since $y$ is expressed in terms of $x$ and $a$, it is not surprising to find that the derivative is also expressed in the same terms. However if we had been asked to integrate

$$\int \frac{1}{\sqrt{a^2-x^2}} dx$$

it would have come as a surprise that that function should include a trigonometrical ratio.

In connection with the calculation of a circle's area it will later be necessary to know the integral of $\int \sqrt{a^2-x^2}\,dx$, in which the integrand is the reciprocal of that discussed above. The existence of square roots in an integrand is always confusing but with a little knowledge of trigonometry the difficulties may often be overcome.

If we substitute $a \sin \theta$ for $x$, then $a^2-x^2$ becomes $a^2-a^2 \sin^2 \theta$ or $a^2(1-\sin^2 \theta)$.

But $\quad 1-\sin^2 \theta = \cos^2 \theta \quad$ (see Appendix II)

$$\therefore\ a^2-x^2 = a^2 \cos^2 \theta$$

This last expression is a perfect square and its square root is easily found to be $a \cos \theta$.

$\int \sqrt{a^2-x^2}\,dx$ may therefore be rewritten as $\int \sqrt{a^2 \cos^2 \theta}\,dx$ or $\int a \cos \theta\,dx$. One further complication remains since this form includes two variables, $x$ and $\theta$. We must therefore express $dx$ in terms of $\theta$. Since we have made

$$x = a \sin \theta$$

then, differentiating with respect to $\theta$

$$\frac{dx}{d\theta} = a \cos \theta$$

whence $\quad dx = a \cos \theta\, d\theta$

The integral may now therefore be re-expressed as

$$\int a \cos \theta . a \cos \theta\, d\theta = \int a^2 \cos^2 \theta\, d\theta = a^2 \int \cos^2 \theta\, d\theta$$

But $\quad \cos^2 \theta = \frac{1}{2}(1+\cos 2\theta) \quad$ (see Appendix II)

whence $\quad a^2 \int \cos^2 \theta\, d\theta = \frac{a^2}{2} \int (1+\cos 2\theta)\, d\theta$

$$= \frac{a^2}{2}(\theta+\tfrac{1}{2} \sin 2\theta)+c$$

But $\sin 2\theta = 2 \sin \theta \cos \theta$ (see Appendix II)

The integral therefore becomes

$$\frac{a^2}{2}(\theta+\sin \theta \cos \theta)+c$$

or, rearranged: $$\underline{\frac{a^2}{2}\theta+\left(\frac{a}{2} \sin \theta\right)(a \cos \theta)+c}$$

We now have to change the variable back from $\theta$ to $x$. We originally substituted

$$x = a \sin \theta$$

$$\therefore \sin \theta = \frac{x}{a} \quad \text{and} \quad \frac{a}{2}\sin \theta = \frac{x}{2}$$

also $$\theta = \sin^{-1} \frac{x}{a}$$

Furthermore $\cos \theta = \sqrt{1-\sin^2 \theta}$ (see Appendix II)

whence $$a \cos \theta = \sqrt{a^2-a^2 \sin^2 \theta}$$
$$= \sqrt{a^2-x^2}$$

Substituting all these equivalents in the underlined integral above gives the result:

$$\int \sqrt{a^2-x^2}\, dx = \frac{a^2}{2}\sin^{-1} \frac{x}{a}+\frac{x}{2}\sqrt{a^2-x^2}+c$$

This is the final result. Its appearance is much less simple than the integrand from which it is obtained. The process by which it is obtained is also a complicated one calling for a working knowledge of trigonometry and the knowledge of how and when to substitute equivalents in the working. By similiar substitution and transformation it is possible to differentiate and to integrate expressions involving other inverse trigonometric ratios. It is beyond the scope of this book to explain them all in detail, but they are listed here for convenience (it being again remembered that $y$ has been restricted to the first quadrant – see page 179).

TRIGONOMETRIC FUNCTIONS

| *Function y* | $\frac{dy}{dx}$ |
|---|---|
| $\sin^{-1} x$ | $\frac{1}{\sqrt{1-x^2}}$ |
| $\cos^{-1} x$ | $-\frac{1}{\sqrt{1-x^2}}$ |
| $\tan^{-1} x$ | $\frac{1}{1+x^2}$ |
| $\cot^{-1} x$ | $-\frac{1}{1+x^2}$ |
| $\sec^{-1} x$ | $\frac{1}{x\sqrt{x^2-1}}$ |
| $\operatorname{cosec}^{-1} x$ | $-\frac{1}{x\sqrt{x^2-1}}$ |

*Chapter 14*

# HYPERBOLIC FUNCTIONS

In addition to the trigonometrical or circular functions there is another important group of functions which are closely related to them. These are the hyperbolic functions which are related to the properties of the hyperbola in much the same way as the circular functions are related to the properties of the circle.

The main functions in this group are the hyperbolic sine, cosine, and tangent which are represented as sinh, cosh, and tanh respectively by the simple expedient of adding an h to the relative circular function. As with the latter, the hyperbolic functions have reciprocals and these are referred to as cosech, sech, and coth respectively.

These functions are mathematically defined as:

$$\sinh x = \tfrac{1}{2}(e^{x}-e^{-x})$$

$$\cosh x = \tfrac{1}{2}(e^{x}+e^{-x})$$

$$\tanh x = \frac{\sinh x}{\cosh x} = \frac{e^{x}-e^{-x}}{e^{x}+e^{-x}} = \frac{e^{2x}-1}{e^{2x}+1}$$

$$\operatorname{cosech} x = \frac{1}{\sinh x} = \frac{2}{e^{x}-e^{-x}}$$

$$\operatorname{sech} x = \frac{1}{\cosh x} = \frac{2}{e^{x}+e^{-x}}$$

$$\coth x = \frac{1}{\tanh x} = \frac{e^{2x}+1}{e^{2x}-1}$$

The close relationship between the hyperbolic and the circular functions may be expressed in the following identities

$$\sinh x = \frac{1}{i} \sin ix$$

$$\cosh x = \cos ix$$

where $i = \sqrt{-1}$. These expressions are included here merely to establish the relationship between the two types of functions. It is not intended to develop these relationships in any examples.

Of more importance to our immediate requirements however is a comparison of formulae by which the functions may be related to each other within their own groups. There are many formulae which relate the circular functions to each other; some of these have already been used here. There are similar formulae for the hyperbolic functions and these indeed are very like those for the circular functions.

In this connection there is a rule, known as Osborne's Rule, which states that, for every connecting formula for circular functions, the corresponding formula connecting hyperbolic functions may be obtained as follows. In place of each circular function in the given formula substitute the corresponding hyperbolic function *provided that* the sign of every product or implied product of two sines is changed. The following examples illustrate the rule, the circular formula being given first in each case.

(i)
$$\cos^2 x+\sin^2 x = 1$$
$$\cosh^2 x-\sinh^2 x = 1$$

(ii)
$$\cos^2 x-\sin^2 x = \cos 2x$$
$$\cosh^2 x+\sinh^2 x = \cosh 2x$$

(iii)
$$\sec^2 x = 1+\tan^2 x$$
$$\operatorname{sech}^2 x = 1-\tanh^2 x$$

The reason for the change in sign of $\tanh^2 x$ is that

$$\tanh^2 x = \frac{\sinh^2 x}{\cosh^2 x}$$

and the product $\tanh^2 x$ is seen to involve the implied squaring of $\sinh x$.

The hyperbolic functions are important in the integration of certain kinds of functions. As one practical example of their uses we may note that the curve of one of the functions ($\cosh x$) gives the general form of the *catenary*. This is the curve formed by a uniform flexible chain or inelastic cable hanging freely but having its ends fixed. Examples of the curve abound in practical applications. The overhead cables of electricity pylons and of telephone wires all hang in catenary form, as also does a skipping rope if it is held firmly by two people and the curve is not distorted by contact with the ground.

The actual shape of the curve formed by the rope will of course depend upon the length of the rope (let this $= a$) and the horizontal distance between the points from which it is suspended (let this $= x$). Then the curve is of the form

$$y = \frac{a}{2}\left(e^{\frac{x}{a}} + e^{-\frac{x}{a}}\right)$$

Because of the similarity between circular and hyperbolic functions it is not surprising that their derivatives should also exhibit similarities. Because the hyperbolic functions look complicated it might be thought that their differentiation would be correspondingly difficult. It will be remembered that the function $e^x$ has a remarkable property in that its derivative is equal to itself. Similarly the derivative of $e^{-x}$ is $-e^{-x}$, the same as the original function but with its sign altered. The functions may therefore be differentiated as follows:

*Hyperbolic sine* $\quad y = \sinh x = \dfrac{e^x - e^{-x}}{2}$

$$\therefore \frac{dy}{dx} = \frac{e^x - (-e^{-x})}{2} = \frac{e^x + e^{-x}}{2} = \underline{\cosh x}$$

*Hyperbolic cosine* $\quad y = \cosh x = \dfrac{e^x + e^{-x}}{2}$

$$\therefore \frac{dy}{dx} = \frac{e^x - e^{-x}}{2} = \underline{\sinh x}$$

*Hyperbolic tangent* $\quad y = \tanh x = \dfrac{\sinh x}{\cosh x}$

Then, by the quotient rule,

$$dy = \frac{\cosh x \cosh x - \sinh x \sinh x}{\cosh^2 x}$$

$$= \frac{\cosh^2 x - \sinh^2 x}{\cosh^2 x}$$

The numerator of this fraction is equal to

$$\left(\frac{e^x + e^{-x}}{2}\right)^2 - \left(\frac{e^x - e^{-x}}{2}\right)^2$$

which may be expanded to

$$\tfrac{1}{4}[(e^{2x} + e^{-2x} + 2) - (e^{2x} + e^{-2x} - 2)] = \tfrac{1}{4}(2+2) = 1$$

Accordingly

$$\frac{dy}{dx} = \frac{1}{\cosh^2 x} = \underline{\operatorname{sech}^2 x}$$

These results should be compared with the results obtained on differentiating the three basic circular functions. The only difference of principle which emerges is that the derivative of the hyperbolic cosine is positive in sign whereas that of the circular cosine is negative.

Inverse hyperbolic functions in some ways correspond to inverse circular functions in character. They may be differentiated by methods similar to those used for the latter and the derivatives in the two groups again offer interesting comparisons.

(i) Let $y = \sinh^{-1} x$

Then $x = \sinh y$

$$\therefore \frac{dx}{dy} = \cosh y \quad \text{and} \quad \frac{dy}{dx} = \frac{1}{\cosh y}$$

But $\cosh^2 y - \sinh^2 y = 1$ Therefore $\cosh y = \sqrt{1+\sinh^2 y}$

$$\therefore \frac{dy}{dx} = \frac{1}{\sqrt{1+\sinh^2 y}} = \frac{1}{\sqrt{1+x^2}}$$

(Since $x = \sinh y$.)

(ii) Let $y = \cosh^{-1} x$

then $x = \cosh y$

$$\therefore \frac{dx}{dy} = \sinh y \quad \text{and} \quad \frac{dy}{dx} = \frac{1}{\sinh y}$$

But $\sinh y = \sqrt{\cosh^2 y - 1}$ and $\cosh^2 y = x^2$

$$\therefore \frac{dy}{dx} = \frac{1}{\sqrt{x^2-1}}$$

(iii) Let $y = \tanh^{-1} x$

then $x = \tanh y$

$$\therefore \frac{dx}{dy} = \operatorname{sech}^2 y \quad \text{and} \quad \frac{dy}{dx} = \frac{1}{\operatorname{sech}^2 y}$$

But $\operatorname{sech}^2 y = 1 - \tanh^2 y$ and $\tanh^2 y = x^2$

$$\therefore \frac{dy}{dx} = \frac{1}{1-x^2}$$

These and the derivatives of the other inverse functions are summarized as follows (where $y$ is restricted to lie in the first quadrant).

| *Function y* | $\frac{dy}{dx}$ |
|---|---|
| $\sinh^{-1} x$ | $\frac{1}{\sqrt{1+x^2}}$ |
| $\cosh^{-1} x$ | $\frac{1}{\sqrt{x^2-1}}$ |
| $\tanh^{-1} x$ | $\frac{1}{1-x^2}$ |
| $\text{cosech}^{-1} x$ | $-\frac{1}{x\sqrt{1+x^2}}$ |
| $\text{sech}^{-1} x$ | $-\frac{1}{x\sqrt{1-x^2}}$ |
| $\coth^{-1} x$ | $-\frac{1}{x^2-1}$ |

## HYPERBOLIC SERIES

Series for the hyperbolic sine and cosine may easily be obtained from the series for $e^x$ and $e^{-x}$.

$$e^x = 1+x+\frac{x^2}{\underline{|2}}+\frac{x^3}{\underline{|3}}+\ldots$$

$$e^{-x} = 1-x+\frac{x^2}{\underline{|2}}-\frac{x^3}{\underline{|3}}+\ldots$$

The hyperbolic sine of $x$ is equal to half the difference between $e^x$ and $e^{-x}$. Thus, subtracting each term of the second equation from the related term in the first equation and then dividing by 2

$$\sinh x = x+\frac{x^3}{\underline{|3}}+\frac{x^5}{\underline{|5}}+\ldots$$

On the other hand, the hyperbolic cosine is equal to half the sum of $e^x$ and $e^{-x}$.

$$\cosh x = 1+\frac{x^2}{\underline{|2}}+\frac{x^4}{\underline{|4}}+\ldots$$

## SUBSTITUTION USES

Many expressions may be integrated by first substituting hyperbolic equivalents. In dealing with the lengths of arcs (Chapter 16) we shall need to know the integral of $\int \sqrt{x^2+a^2}\,dx$. This integration is now demonstrated as an example of the use of hyperbolic functions.

First, let $$x = a \sinh z \qquad \text{(i)}$$

Then, also $$z = \sinh^{-1} \frac{x}{a} \qquad \text{(ii)}$$

and $$\sinh z = \frac{x}{a} \qquad \text{(iii)}$$

If we differentiate (i) we obtain:

$$\frac{dx}{dz} = a \cosh z$$

$$\therefore\ dx = a \cosh z\, dz \qquad \text{(iv)}$$

Next, we have from (iii):

$$\sinh z = \frac{x}{a}$$

But $\cosh^2 z - \sinh^2 z = 1$ (see page 185) (v)

$$\therefore\ \cosh z = \sqrt{1+\sinh^2 z}$$

$$= \sqrt{1+\left(\frac{x}{a}\right)^2} \quad \textit{from (iii)}$$

$$= \frac{1}{a}\sqrt{a^2+x^2} \qquad \text{(vi)}$$

These equations provide us with the tools required for integration. Substituting from (i) in the original expression gives:

$$\int \sqrt{x^2+a^2}\,dx = \int \sqrt{a^2 \sinh^2 z + a^2}\,dx$$

$$= \int a\sqrt{\sinh^2 z+1}\,.\,a \cosh z\, dz \quad \textit{from (iv)}$$

$$= \int a \cosh z\,.\,a \cosh z\, dz \quad \textit{from (v)}$$

$$= \int a^2 \cosh^2 z\, dz$$

But $\cosh^2 z = \frac{1}{2}(\cosh 2z+1)$ *see Appendix II*

$$\therefore \int \sqrt{x^2+a^2}\, dx = \frac{a^2}{2}\int (\cosh 2z+1)\, dz$$

$$= \frac{a^2}{2}(\tfrac{1}{2}\sinh 2z+z)+c$$ *on integrating*

But $\sinh 2z = 2 \sinh z \cosh z$ *see Appendix II*

$$\therefore \int \sqrt{x^2+a^2}\, dx = \frac{a^2}{4}(2 \sinh z \cosh z)+\frac{a^2 z}{2}+c$$

$$= \tfrac{1}{2}(a \sinh z)(a \cosh z)+\frac{a^2 z}{2}+c$$

But $a \sinh z = x$ *from (i)*

$a \cosh z = \sqrt{a^2+x^2}$ *from (vi)*

$z = \sinh^{-1} \frac{x}{a}$ *from (ii)*

$= \log\left(\frac{x+\sqrt{x^2+a^2}}{a}\right)$ *see Appendix IV*

Substituting all these equivalents now gives:

$$\int \sqrt{x^2+a^2}\, dx = \tfrac{1}{2}x\sqrt{x^2+a^2}+\frac{a^2}{2}\left(\log \frac{x+\sqrt{x^2+a^2}}{a}\right)+c$$

But $\log\left(\frac{x+\sqrt{x^2+a^2}}{a}\right) = \log (x+\sqrt{x^2+a^2})-\log a$

Finally,

$$\int \sqrt{x^2+a^2}\, dx = \tfrac{1}{2}x\sqrt{x^2+a^2}+\frac{a^2}{2}\log (x+\sqrt{x^2+a^2})-\frac{a^2}{2}\log a+c$$

The term $\frac{a^2}{2}\log a$ is a constant which may be regarded as forming part of the constant of integration. The standard form of this integral therefore omits this term (see note at the end of Appendix IV).

## *Chapter 15*

# PLANE AREAS

One of the main spurs to action which resulted in the development of the integral calculus was the necessity for finding a means of calculating the areas and volumes of curved surfaces or solids.

It has already been shown that it is possible to calculate the area 'under a curve' by integration. This principle may be extended to the calculation of plane surface areas within curved boundaries. If the shapes of these boundaries are the same as the shapes of curves representing functions, then the properties of these functions may be used in calculations involving the boundaries.

We may draw the curve of a function on squared paper and then cut out the shape enclosed by the curve, the ordinate, and the two axes. If the function is $y = x^2$ and we draw the part of the curve between $x = 0$ and $x = 10$, then the shape of the piece of paper cut out will be as shown in Fig. 37.

Before cutting the shape out we could have calculated its area, since it is the same as the area under the curve. Since $y = x^2$ then by integration the area under the curve is $\frac{1}{3}x^3$ square units. This is clearly also the area of the piece of paper cut out. The piece of paper is a distinct and separate entity, but we can calculate its area because it can be fitted to the curve. One calculation gives both areas.

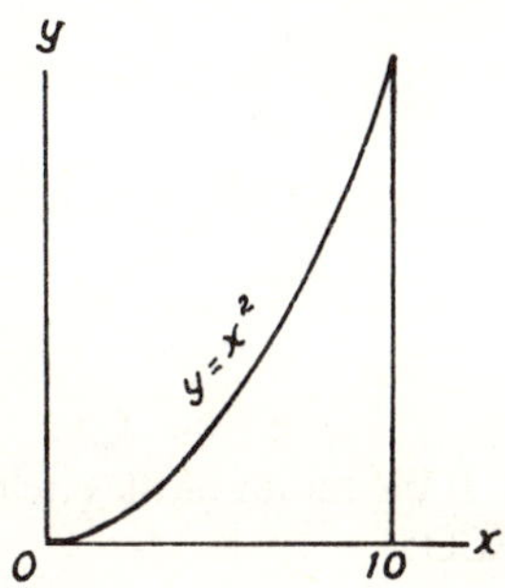


FIG. 37

### AREA OF A CIRCLE

In general, therefore, if we wish to calculate the area of a curved shape we must first identify the function of the curve to which the shape's curved boundary may be fitted.

To calculate the area of a circle we need to find a function whose curve is itself a circle.

Fig. 38 shows a circle centred at O, the origin of the co-ordinate axes. E is any point on the circumference and EC is drawn at right angles to the $x$ axis. We concentrate on the right upper quadrant (or first quadrant) of the circle which lies between the two axes and regard the arc VEW as the curve of a function (for positive values of $x$ and $y$) drawn relative to the two axes. Then EC is the ordinate of this curve at point E. EC thus represents the value of $y$ relative to the value of $x$ at point C.

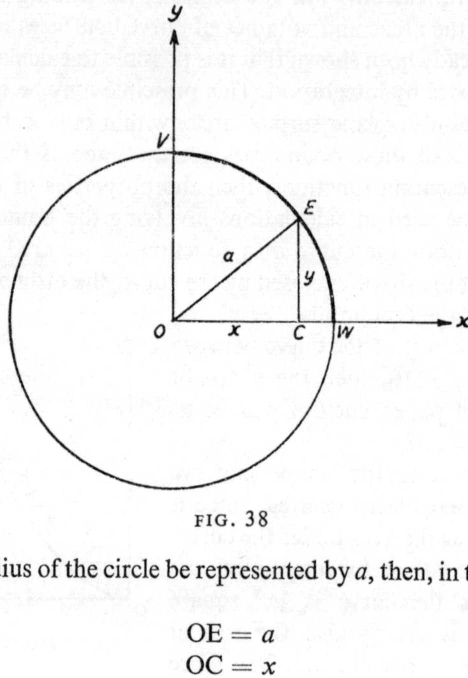


FIG. 38

If the radius of the circle be represented by $a$, then, in the triangle ECO:

$$\text{OE} = a$$
$$\text{OC} = x$$
$$\text{EC} = y$$

But EC is perpendicular to OC, so that ECO is a right-angled triangle. Therefore

$$x^2+y^2 = a^2$$

or

$$y = \sqrt{a^2-x^2}$$

Now E is any point on the circle, and the relationship between the sides of the triangle ECO is the same for any and every point on the circle. Accordingly the circle describes a curve which represents the function $y = \sqrt{a^2-x^2}$. The methods of integration may now be applied to this function to calculate the area under the curve and this will tell us the area of the quadrant of the circle. It is only necessary then to multiply this area by four to calculate the area of the whole circle.

The area of the quadrant has for its limits the values of $x$ at O and at W. At O; $x = 0$. At W; $x = a$, since OW is a radius of the circle. We therefore integrate the function between these two limits:

$$\text{Area} = \int_0^a \sqrt{a^2-x^2}\,dx$$

Here we can recognize an integrand, the integral of which was shown in Chapter 13 to be as follows:

$$\int \sqrt{a^2-x^2}\,dx = \frac{a^2}{2}\sin^{-1}\frac{x}{a}+\tfrac{1}{2}x\sqrt{a^2-x^2}+c$$

We can ignore the constant of integration because we are integrating between limits. (In dealing with the inverse function it should be remembered that we are restricting our attention to the first quadrant of the circle.)

*When* $x = 0$, the value of $\sin^{-1}\frac{x}{a} = \sin^{-1} 0 = 0$. Substituting these values in the integral makes the integral also $= 0$.

*When* $x = a$, the value of $\sin^{-1}\frac{x}{a} = \sin^{-1} 1$. This represents an angle the sine of which is 1. The angle is therefore 90° or, in circular measure, $\frac{\pi}{2}$ radians. That is $\sin^{-1} 1 = \frac{\pi}{2}$. Substituting this value in the integral and also substituting $a$ for $x$, transforms the integral to:

$$\frac{a^2}{2}\cdot\frac{\pi}{2}+\tfrac{1}{2}a\sqrt{a^2-a^2} = \frac{\pi a^2}{4}+0 = \frac{\pi a^2}{4}$$

Consequently the area between the limits $x = 0$ and $x = a$ is

$$\frac{\pi a^2}{4}-0 = \frac{\pi a^2}{4}$$

But this is a quadrant of the circle. The total area of the circle is therefore

$$\frac{\pi a^2}{4} \times 4 = \pi a^2$$

The same method may be employed to calculate the area of any symmetrical regular plane figure by regarding its curve as being generated by a function of $x$. Before proceeding to consider further examples of this type, it is instructive to discuss other methods for calculating the area of a circle; methods which depend for their validity upon the notion of limits but which are not directly linked to the curves of functions.

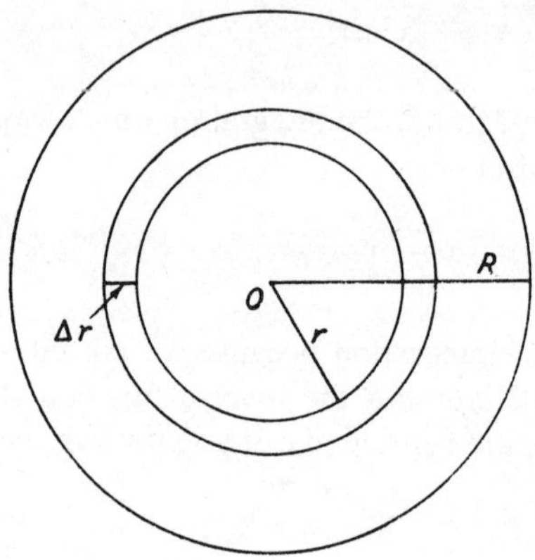

FIG. 39

One simple method follows. Draw a circle with centre O and radius $R$ as in Fig. 39. Draw a small ring of width $\Delta r$, placed at a distance $r$ from the centre. As this ring is made progressively smaller (i.e. narrower), so its width $\Delta r$ will approach zero and the length of the outer circumference of the ring will approach equality with the length of the inner circumference of the ring; that is, it will approach $2\pi r$.

The area of the ring is therefore approximately equal to

$$2\pi r\, \Delta r$$

(the approximation being as close as we like by making $\Delta r$ sufficiently small). The whole circle is regarded as being made up of an indefinite number of such rings and the area of the circle may therefore be calculated by taking the limiting value of the sum of the areas of all the rings taken together $= \int 2\pi r\, dr$ between the limits of $r = 0$ and $r = R$.

$$A = \int dA = \int_0^R 2\pi r\, dr = 2\pi \int_0^R r\, dr$$

$$= 2\pi \left[\frac{1}{2}r^2\right]_0^R$$

When $r = R$ then $\frac{1}{2}r^2 = \frac{1}{2}R^2$

When $r = 0$ then $\frac{1}{2}r^2 = 0$

$$\therefore\ A = 2\pi[\tfrac{1}{2}R^2 - 0] = \pi R^2$$

Here is another method. A straight line OB is rotated about point O so that point B will describe the circle in Fig. 40. Let OB $= a$. If it is assumed that the line OB has rotated from its position at OA and in doing so has described the angle $\theta$, then we may similarly let OB rotate to OC through an increment of angular measure equal to $\Delta\theta$. The arc subtended by $\Delta\theta$ is equal to $a\ \Delta\theta$ (see explanation of circular measure in Chapter 13).

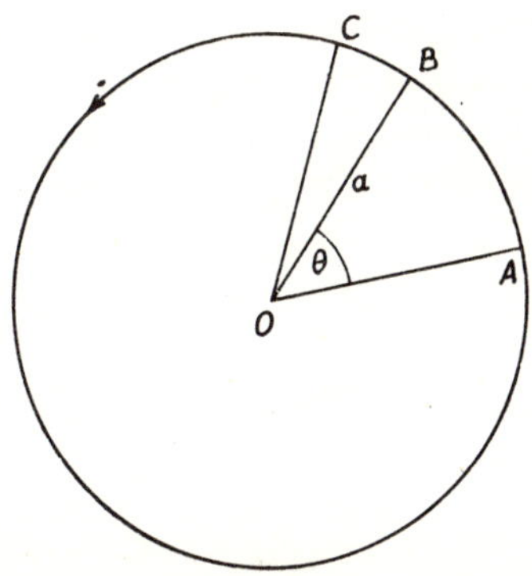


FIG. 40

If we were to draw the tangent to the circle at C and extend OB to meet this tangent at X, we would form a right-angled triangle OCX, the area of which would be:

$$(\tfrac{1}{2}\text{CX})(\text{height of triangle}) = (\tfrac{1}{2}\text{CX})(a)$$

(This triangle is not shown in the diagram because of constructional difficulties.)

As we make $\Delta\theta$ smaller, the length of arc CB will approach the length of CX as its limit. But the length of the arc is $a\ \Delta\theta$ whence CX will be approximately equal to $a\ \Delta\theta$, and the area of the triangle will be approximately equal to $\frac{1}{2}(a\ \Delta\theta)a = \frac{1}{2}a^2\ \Delta\theta$ (the approximation being as close as we like by making $\Delta\theta$ sufficiently small).

The area of the circle is equal to the limiting value of the sum of the areas of all these triangles between the limits of $\theta = 0$ and $\theta = 2\pi$. That is,

$$A = \int_0^{2\pi} \tfrac{1}{2}a^2\, d\theta$$

$$= \frac{a^2}{2}\int_0^{2\pi} d\theta \qquad \text{(since } a \text{ is constant)}$$

$$= \frac{a^2}{2}\Big[\theta\Big]_0^{2\pi}$$

$$= \left(\frac{a^2}{2}\right)(2\pi) = \pi a^2$$

## POLAR CO-ORDINATES

This last method of calculating the area of a circle serves to introduce the idea of polar co-ordinates. These locate the position of a point in a plane just as do rectangular co-ordinates, but they employ different kinds of measurements. Polar co-ordinates locate the position of a point by its distance from a fixed point, called the *pole* or *origin*, and the point's angular direction from a given fixed line drawn through the pole.

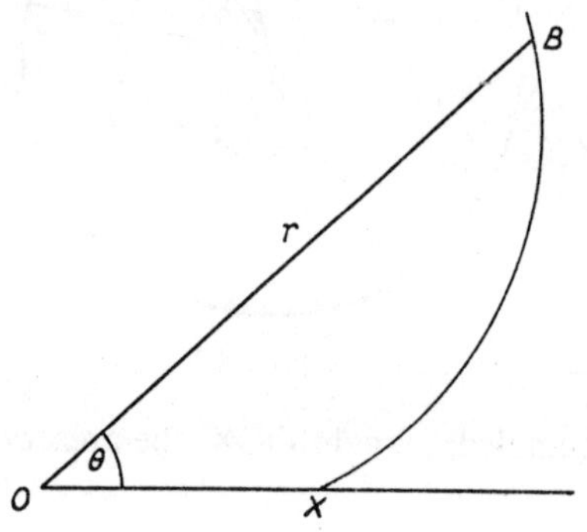

FIG. 41

In Fig. 41, the line OX is fixed, the pole being at O. The line OB ($= r$) which joins the point B and the pole is called the *radius vector*, and $\theta$ which gives the angular direction of the vector is called the *vectorial angle*. It will be seen that, given OX, the co-ordinates $r$ and $\theta$ are sufficient to locate the position of B.

The relation $y = f(x)$ may be represented as a curve drawn in relation to rectangular co-ordinates. Similarly the relation $r = f(\theta)$ may be represented as a curve drawn relative to polar co-ordinates. Fig. 41 illustrates a curve of this type. How can we calculate the area of the sector of the curve between OX and OB?

If the curve had been a circle then we know from the last example that the area of the sector would have been

$$A = \frac{1}{2}\int_0^\theta r^2\,d\theta$$

which is simplified to

$$A = \frac{1}{2}r^2\int_0^\theta d\theta \qquad \text{since } r \text{ is constant}$$

The curve in Fig. 41 is obviously not a circle and $r$ is therefore not constant. The area of the sector XOB may however be regarded as consisting of indefinitely small sectors each of which may in turn be regarded as the sector of a circle. But they are not sectors of the *same* circle. For each separate sector there is a separate and different circle from all the others.

The area of each such sector is approximately $\frac{1}{2}r^2\,\Delta\theta$ and the limiting value of the sum of all the areas is

$$\frac{1}{2}\int_0^\theta r^2\,d\theta \qquad \text{or} \qquad \frac{1}{2}\int_0^\theta f(\theta)^2\,d\theta$$

between the specified limits. As $r$ is not constant for this curve we cannot transfer $r$ to the left side of the integral sign.

Polar co-ordinates are often used in connection with problems involving the properties of a circle or which may be transformed into such problems. The use of polar co-ordinates in the calculus generally increases in the more advanced stages of the subject. They are not however necessary for the elementary stages and, for the most part, we shall continue to use rectangular co-ordinates.

## SIGN OF AN AREA

We now revert to the first method of calculating the area of a circle by identifying a part of it with the area under the curve of a function drawn relative to $x$ and $y$ co-ordinate axes.

We first calculated the area of a quadrant of the circle and then quadrupled this to obtain the area of the whole circle. Why was it not possible to integrate the whole area in one calculation? This raises a very important point which requires some explanation.

First consider a very different function:

$$y = 4x^3 - 12x^2 + 8x$$

the curve of which is shown in Fig. 42. We are asked to calculate the area under the curve between the limits of $x = 0$ and $x = 2$.

If we integrate $$\int_0^2 (4x^3-12x^2+8x)\,dx$$

we obtain $$A = \Big[x^4-4x^3+4x^2\Big]_0^2$$

when $x = 0$ the integral $A = 0$
when $x = 2$ ,, ,, $A = 0$

so that the value of the area between these two limits would be shown as equal to $0-0 = 0$. But the figure clearly shows that

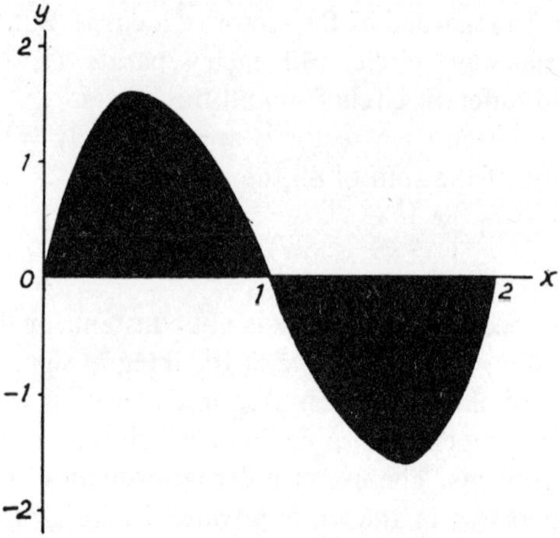


FIG. 42

a piece of paper cut out in the shape of the shaded part would have an area which could not be equal to zero.

In calculating the area 'under a curve' it is essential to bear in mind exactly what this term means. Another glance at Fig. 42 will show that half the area is above the $x$ axis, while the other half is below the axis. The area of the half above the axis is positive while the area below the axis is negative. The separate areas of the two parts of the shape are in fact equal but of opposite signs (for example $+A$ and $-A$) so that when they are added together we obtain $+A+(-A) = 0$.

Here we must distinguish between the algebraic sum of the two areas. This sum is indeed equal to zero as shown. But this makes nonsense of the practical problem of calculating the physical area of a shape whose curved boundary coincides with the curve of the function. This problem does not of course arise with every function.

The function we have chosen may be factorized

$$y = 4x^3-12x^2+8x = 4x(x-1)(x-2)$$

and it will be apparent from this factorization that $y = 0$ when $x = 0$, 1, or 2. The curve must cut the $x$ axis at each of these values and part of the curve must lie below the axis.

The area below the $x$ axis is negative only because of its position relative to that axis. Yet for the purpose of calculating the area of a physical shape it is clear that the whole of this area is positive irrespective of its situation. The fact that we may place such a shape so that its curved boundaries coincide with the curve of the function and that part of the shape must therefore be placed below the $x$ axis does not alter the character of the shape. The $x$ axis has no meaning in relation to the physical shape.

In applying integration methods to practical shape problems, therefore, we must rule out the possibility of negative areas. But ordinary integration between the two specified limits does not ignore this possibility since the method of integration adopted is algebraic and will therefore produce the algebraic area under the curve of the function. While noting that the curve of the shape is identical with the curve of the function we must also distinguish differences between the two curves. The physical shape is not the same as the graph any more than a gold strip is the same as a piece of wood merely because they share some common dimension.

To calculate the combined area of the two parts of the shape we must ignore the negative sign of the area which coincides with the area of the curve below the axis. In order to do this however the areas of the two parts must be calculated separately. If any area is mathematically negative its sign must first be changed to positive before the separate areas are added together. Thus the area of the shape, as distinct from the area under the curve, is

$$A = \int_0^1 (4x^3-12x^2+8x)\,dx - \int_1^2 (4x^3-12x^2+8x)\,dx$$

$$= \Big[x^4-4x^3+4x^2\Big]_0^1 - \Big[x^4-4x^3+4x^2\Big]_1^2$$

(The sign of the second integral has been changed for the reason given above.)

The first integral

$$= (1-4+4)-0 \qquad = \quad 1$$

The second integral

$$= (16-32+16)-(1-4+4) = -1$$

$$\text{Total area} = 1-(-1) = \quad 2 \quad \text{(square units)}$$

It is now clearer why, in calculating the area of a circle, we first calculated the area of a quadrant. If we had tried to calculate the whole area of the circle in one integration this area would have been shown to be equal to zero. This is because one half of the circle is above the $x$ axis while the other half is below it.

In addition to this the curve described by the circumference of a circle has another confusing property; the property which in fact results in half of the curve falling below the $x$ axis. The equation of this curve is

$$y = \pm\sqrt{a^2-x^2}$$

The square root of a number may be either positive or negative. Thus for every value of $x$ there is a positive and a negative value for $y$. If $a = 5$ then, when $x = 4$, so $a^2-x^2 = 9$, and the square root of this is $\pm 3$. By confining our attention to the upper right quadrant of the circle we remove this confusion because we know from the properties of the co-ordinate system that all values of $y$ are positive in this quadrant.

To summarize therefore it will be seen that we must not attempt to calculate the area under a curve or, by extension, the area of a curved plane figure without a knowledge of the properties of the curve relative to the co-ordinate axes. The shape whose area is to be calculated must first be inspected to reveal whether any part of it would by ordinary integration methods be treated as if it were negative.

## AREA OF AN ELLIPSE

Another interesting curve is that described by the perimeter of an ellipse. The ellipse is a most important curve in nature. For centuries it was thought that the planets moved along circular paths around the sun. Kepler, who has already been mentioned as an early calculus pioneer, was to show that in fact the orbits of the planets are elliptical and not circular. Newton later proved why this should be so and the principles disclosed are now used in planning the orbits of space ships.

Fig. 43 represents an ellipse the centre of which is at O, the

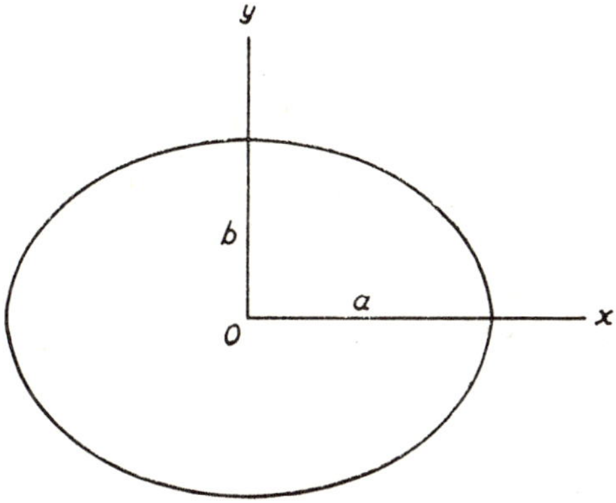


FIG. 43

origin of the axes. The equation of this ellipse is

$$\frac{x^2}{a^2}+\frac{y^2}{b^2} = 1$$

whence

$$y = \pm\frac{b}{a}\sqrt{a^2-x^2}$$

We again concentrate on the right upper quadrant. The area of a small strip of width $\Delta x$ is approximately

$$y\,\Delta x = \frac{b}{a}\sqrt{a^2-x^2}\,\Delta x$$

(the approximation being as close as we like by making $\Delta x$ sufficiently small). The area of the quadrant is therefore the limiting value of the sum of all such elements of area

$$= \int_0^a \frac{b}{a}\sqrt{a^2-x^2}\,dx = \frac{b}{a}\int_0^a \sqrt{a^2-x^2}\,dx$$

It has been shown that the area of a quadrant of a circle with radius $a$ is equal to $\int_0^a \sqrt{a^2-x^2}\,dx$ and that the total of the circle's area is equal to $\pi a^2$. It is now seen that the area of a quadrant of an ellipse is $\frac{b}{a}$ times the area of a quadrant of a circle. Consequently the area of the ellipse is equal to $\frac{b}{a}\pi a^2$ or $\pi ab$.

## IRREGULAR SHAPES

The circle and the ellipse are shapes described by curves representing functions. Can the calculus provide a method for calculating the area of irregular shapes (such as that in Fig. 44) which are

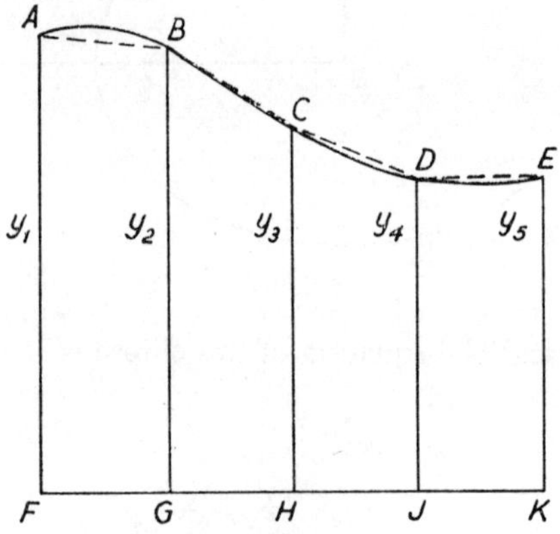


FIG. 44

bounded on one or more sides by curves which are not identical with curves of any known functions? It should be made clear at once that there is no known method for calculating the area of every shape of this type exactly. But the calculus provides or at least justifies a method which, although still approximate, nevertheless gives an approximation which is more nearly accurate than that given by practical methods such as counting the number of squares on graph paper covered by the shape.

Reference is now made to the shape AEKF in Fig. 44. AF and EK are perpendicular to FK. Perpendicular lines are also drawn at points G, H, and J to cut the curved upper boundary at B, C, and D respectively, and so that the measurements of FG, GH, HJ, and JK are equal to each other. Next each pair of consecutive points on the curve are joined by a straight line; that is by lines AB, BC, etc.

Then each strip such as ABGF is a trapezium and its area may be calculated by simple geometry. If the base FG $= a$; and if AF and BG $= y_1$ and $y_2$ respectively then the area of ABGF trapezium $= \frac{a}{2}(y_1+y_2)$. The addition of the areas of all such trapeziums approximates to the area of the shape bounded by the curve.

The application of these approximations is known as the trapezoidal rule. The greater the number of trapeziums constructed within the shape so also the more closely will their total area approximate to the area of the original shape.

Making the trapeziums smaller and smaller is the kind of process with which we have become familiar in the calculus. We cannot apply ordinary integration methods to the calculation of the area of the shape, since we cannot treat the curved boundary as having been described by a curve of a function. We can however obtain a better approximation by replacing the pairs of straight lines like AB and BC by arcs of suitable regular curves which approximate more closely to the shapes of the curve between the three consecutive points such as A, B, and C.

We therefore assume that the part of the curve joining these three points is an arc of a parabola, the general equation of which is

$$y = a+bx+cx^2$$

and from this we calculate the area of the shape ACHF.

We now have to identify the part of the curve ABC with the arc of the parabola and for this we must place the shape ACHF relative to the axes of a co-ordinate system. For convenience we assume that the point G coincides with the origin of such a system. The points A, B, and C then coincide with the values of $y$ relative to the values of $x$ at points F, G, and H.

Now FG $=$ GH. Let each of these be represented by the letter $l$. Since G is at the origin of the $x$ axis then at that point $x = 0$. At point F the value of $x = -l$; while at point H the value of $x = +l$. Then the $y$ ordinates at the respective points will, by substituting these values in the above equation, be:

$$\begin{aligned} AF &= y_1 \quad \text{when } x = -l. \text{ Hence } y_1 = a-bl+cl^2 &\text{(i)}\\ BG &= y_2 \quad \text{,,} \quad x = 0. \text{ Hence } y_2 = a &\text{(ii)}\\ CH &= y_3 \quad \text{,,} \quad x = +l. \text{ Hence } y_3 = a+bl+cl^2 &\text{(iii)} \end{aligned}$$

Adding (i) and (iii)

$$y_1+y_3 = 2a+2cl^2$$

But from (ii) $\quad y_2 = a \quad$ (iv)

$$\therefore 2y_2 = 2a$$

$$\therefore y_1+y_3 = 2y_2+2cl^2$$

$$\therefore cl^2 = \tfrac{1}{2}(y_1+y_3-2y_2) \qquad \text{(v)}$$

To integrate the area of ACHF we must calculate the value of

$$\int_{-l}^{+l}(a+bx+cx^2)\,dx$$

$$= \left[ax+\frac{1}{2}bx^2+\frac{1}{3}cx^3\right]_{-l}^{+l}$$

When $\quad x = +l, \quad$ the integral $= al+\tfrac{1}{2}bl^2+\tfrac{1}{3}cl^2$

When $\quad x = -l \quad$ ,, $\quad$ ,, $\quad = -al+\tfrac{1}{2}bl^2-\tfrac{1}{3}cl^2$

Changing the sign of the second value and adding it to the first

$$\text{Area ACHF} = 2al+\tfrac{2}{3}cl^3$$

$$= 2l(a+\tfrac{1}{3}cl^2)$$

But from (iv) and (v) above

$$a = y_2$$

$$cl^2 = \tfrac{1}{2}(y_1+y_3-2y_2)$$

$$\therefore \text{Area} = 2l(a+\tfrac{1}{3}cl^2)$$

$$= 2l\left(y_2+\frac{y_1+y_3-2y}{6}\right)$$

$$= l\left(\frac{y_1+4y_2+y_3}{3}\right)$$

Similarly the area of CEKH

$$= l\left(\frac{y_3+4y_4+y_5}{3}\right)$$

The total area of the whole shape is therefore

$$l\left(\frac{y_1+4y_2+y_3}{3}\right)+l\left(\frac{y_3+4y_4+y_5}{3}\right)$$

$$=\frac{l}{3}\left[(y_1+y_5)+2(y_3)+4(y_2+y_4)\right]$$

This formula is shorthand for the following method of calculating the area:

1. Add (i) Sum of first and last ordinates
   and (ii) Twice the sum of the odd-numbered ordinates (other than the first and the last)
   and (iii) Four times the sum of the even-numbered ordinates
2. Multiply the grand total obtained by one-third of the distance between consecutive ordinates.

This method is known as *Simpson's Rule.*

We must be clear as to what we have done here. For each strip of area we have assumed that its curved boundary approximates very closely to the arc of a curve of a function and we have based our calculations on this assumption. For each different strip the respective arc is assumed to be part of a different parabola. When we add the areas together and so 'reconstruct' the original curved boundary of the whole shape we do not thereby construct a parabola. All we have in theory is an irregular curve which is composed of a number of arcs from different parabolas.

In order to use this rule it is necessary to construct an odd number of ordinates and the closer they are together (that is, the smaller the value of $l$) so the more closely will the result of the calculation approximate to the true area of the shape. It is again stressed that the result can only be approximate and that other errors may creep in because of the limitations involved in measuring the lengths of the ordinates accurately.

The reader who would like to check the efficiency of this method is recommended to employ it to measure the area of a regular figure such as a circle since the results achieved may be checked by other methods. If his measurements are accurate, he will find that the result involves an error of less than one per cent, by taking eleven or more ordinates.

# Chapter 16

# VOLUMES AND SURFACE AREAS

The integral calculus has another very important application in the measurement of volumes and surface areas of *solids of revolution.* This is a mathematical term for solids which may be regarded as having been marked out in space by the rotation of a curve about an axis. Thus, if a semi-circle is rotated about its diameter it will 'sweep out' the volume of a sphere.

## VOLUME OF A SPHERE

Fig. 45 shows a circle with its centre at O, the origin of a system of co-ordinates. This circle represents the cross-section of a sphere the

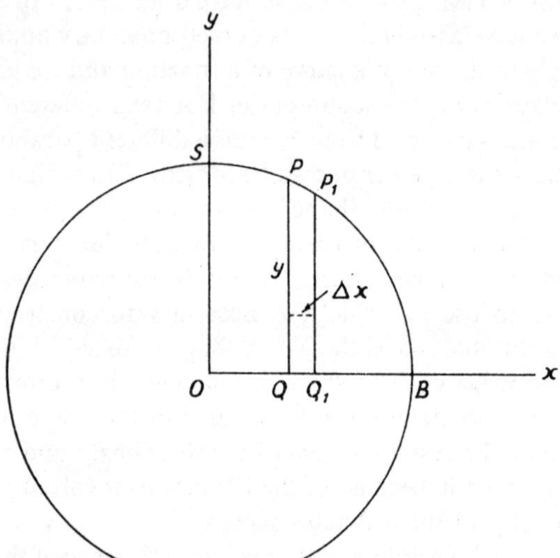


FIG. 45

volume of which we wish to calculate. If we rotate the quadrant OSB about the $x$ axis it will sweep out the volume of half the sphere. If we can calculate this volume then we need only double it to calculate the volume of the whole sphere. This again enables us to concentrate our mathematics on just one quadrant.

First we consider what volume is swept out by the shaded strip $PP_1QQ_1$. Let the width of this strip be $\Delta x$. That is $QQ_1 = \Delta x$. If the ordinate at P is $y$ then, as $x$ grows to $x+\Delta x$ at $Q_1$, so $y$ grows to $y+\Delta y$ at $P_1$. When the strip is rotated about the $x$ axis it will therefore sweep out the volume of a circular disc, the circular sides of which have $y$ and $y+\Delta y$ for their respective radii. The width of the disc is $\Delta x$. Consequently the volume of the disc lies between $\pi y^2 \Delta x$ and $\pi(y+\Delta y)^2 \Delta x$.

If we now move $P_1$ indefinitely close to P so that $\Delta x$ becomes indefinitely small, then as $\Delta x$ approaches zero, the length of $P_1Q_1$ approaches equality with the length of PQ. Accordingly the volume will now be approximately equal to $\pi y^2 \Delta x$, the approximation being as close as we like by taking $\Delta x$ sufficiently small.

The volume of the hemisphere swept out by the quadrant is the limiting value of the sum of the volumes swept out by all such strips between the values $x = 0$ and $x = a$. The total volume is therefore calculated by integrating $\int \pi y^2\, dx$ between these limits. The equation of the circle is $y = \sqrt{a^2-x^2}$ and the calculation is as follows:

$$\begin{aligned}\text{Volume} &= \int_0^a \pi y^2\, dx \\ &= \pi \int_0^a (a^2-x^2)\, dx \\ &= \pi \Big[ a^2x-\tfrac{1}{3}x^3 \Big]_0^a \\ &= \pi(a^3-\tfrac{1}{3}a^3) = \tfrac{2}{3}\pi a^3\end{aligned}$$

This is the volume of the hemisphere and accordingly the volume of the sphere is twice as great

$$= \tfrac{4}{3}\pi a^3$$

The same principles may be applied to all other regular solids of revolution irrespective of their actual shape. The disc of volume swept out by the element of area under the curve will always be circular whatever the shape of the curve. This may at first seem a little odd, yet it is implicit in the rotation of the curve about its axis. To rotate the curve means to turn it about its axis so that the whole curve describes a lateral circular motion.

If a square piece of wood has a rod fixed through its centre and is then rotated about the rod then every point on it will describe a circular movement even though the object rotated is not itself

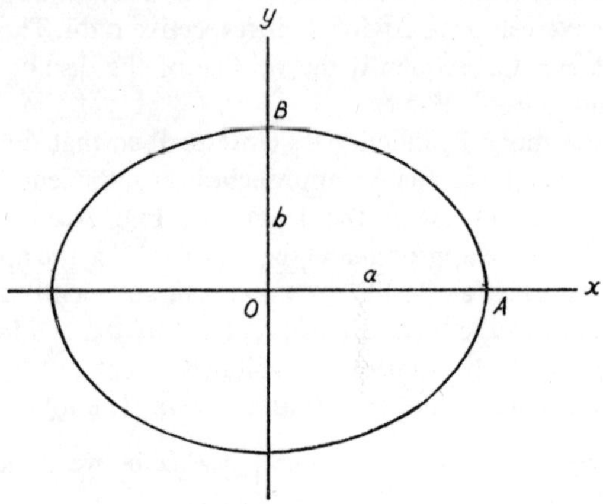


FIG. 46

circular. In the same way, the shape of a curve has no effect upon the shape of the discs of volume swept out by the elements of area. What it does do of course is to govern the lengths of the radii of the different discs since the limiting value of each radius is necessarily a $y$ ordinate of the curve at the respective point.

The fact that each disc of volume swept out is circular is of immense importance since its volume formula must therefore always be $\pi y^2\, dx$ whatever the shape of the solid. The integration of these elements of volume then gives an expression for the total volume of a solid. That is

$$\text{Volume} = \int \pi y^2\, dx$$

This holds for any regular solid. To calculate its volume all that we need do is to substitute the function of $x$ which is represented by $y$ in the generating curve.

## VOLUME OF AN ELLIPSOID

This principle is now applied to the calculation of the volume of an ellipsoid – the solid swept out by half an ellipse. This is represented in Fig. 46.

If the quadrant OBA is rotated about the $x$ axis it will generate one half of the ellipsoid. The equation of the ellipse is

$$\frac{x^2}{a^2}+\frac{y^2}{b^2} = 1$$

where $2a$ = length of major axis; and $2b$ = length of the minor axis.

So $$y^2 = \frac{b^2}{a^2}(a^2-x^2)$$

The volume swept out by the rotated element of area is approximately equal to $\pi y^2 \Delta x$ and the total volume swept out by the quadrant (which has for its limits $x = 0$ and $x = a$)

$$= \int_0^a \pi y^2 \, dx$$

Substituting for $y^2$, this becomes

$$\text{Volume} = \int_0^a \pi\left(\frac{b^2}{a^2}\right)(a^2-x^2)\, dx$$

$$= \frac{\pi b^2}{a^2}\int_0^a (a^2-x^2)\, dx$$

$$= \frac{\pi b^2}{a^2}\left[a^2x-\tfrac{1}{3}x^3\right]_0^a$$

$$= \frac{\pi b^2}{a^2}(a^3-\tfrac{1}{3}a^3)$$

$$= \tfrac{2}{3}\pi ab^2$$

This is the volume of half of the ellipsoid. The volume of the whole ellipsoid is therefore twice as great. Thus the volume of an ellipsoid

$$= \tfrac{4}{3}\pi ab^2$$

## VOLUME OF A CONE

The same principle is now applied to the calculation of the volume of a circular cone.

The cone is generated by the rotation of a right-angled triangle about one of the sides containing the right angle (Fig. 47). The radius of the base of the cone is $r$ (= AM) and the height of the cone is $a$ (= OM). If the side OA is regarded as the curve of a function, its equation will be of the form $y = mx$ since the curve is a straight line.

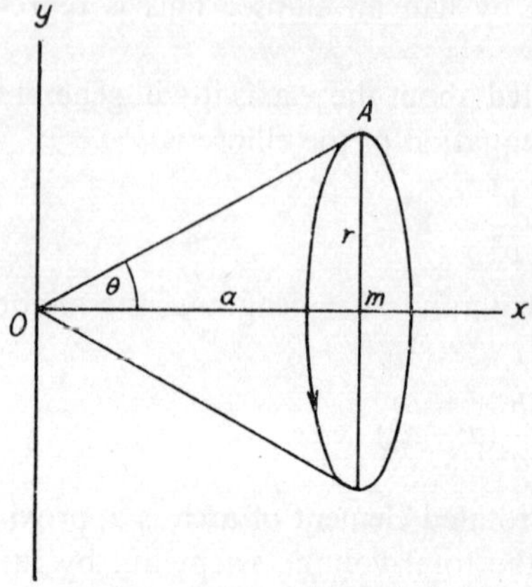


FIG. 47

If the angle MOA = $\theta$, then the derivative of the function is equal to tan $\theta$. The derivative of the function $y = mx$ however is $m$. Consequently

$$m = \tan \theta$$

In its turn, tan $\theta$ is equal to the ratio of side AM to side OM, and this ratio is $\frac{r}{a}$. Thus

$$m = \frac{r}{a}$$

We now proceed to calculate the volume of the cone. Using the general formula, the volume is equal to

$$\begin{aligned} V &= \int_0^a \pi y^2\, dx \\ &= \int_0^a \pi (mx)^2\, dx \\ &= \pi m^2 \int_0^a x^2\, dx \\ &= \pi m^2 \left[\frac{x^3}{3}\right]_0^a = \pi m^2 \left(\frac{a^3}{3}\right) \quad \text{or} \quad \tfrac{1}{3}\pi m^2 a^3 \end{aligned}$$

But $$m = \frac{r}{a}$$

Therefore $$\tfrac{1}{3}\pi m^2 a^3 = \tfrac{1}{3}\pi r^2 a$$

In other words, the volume of the cone is equal to one-third of the height multiplied by the area of the base.

The above methods apply to the calculation of volumes of regular solids only. They cannot be used in reference to irregular solids. Nevertheless the method of calculating the areas of irregular plane figures by Simpson's rule may be extended to calculate the volumes of irregular solids. All that is required for this is the ability to measure the areas of cross-sections of the solid at equal intervals along its axis. The values of these areas may then be plotted as ordinates of an irregular curve such as that shown in Fig. 44. Simpson's rule may then be applied to calculate the area under this curve and this will in fact also give the volume of the solid.

Thus if in Fig. 44 the ordinates represented areas of cross-sections then multiplying each of these in the way shown by a specified distance along the axis (that is, between the cross-sections) would give a numerical answer in cubic units representing the approximate value of the volume.

## LENGTH OF ARCS

The integral calculus also provides a method for calculating the surface area of a regular solid. It is first necessary however to consider the measurement of lengths of arcs of regular curves, since this dimension will be required in calculating the surface area.

In Fig. 48, let AB be the arc of a curve of the function $y = f(x)$. E and F are two points on the curve and they have been joined by a straight line EF which is the chord of the arc between them. In the usual notation EG $= \Delta x$ and FG $= \Delta y$. Similarly the part of the curve between E and F is equal to $\Delta s$, where $s$ measures length of curve.

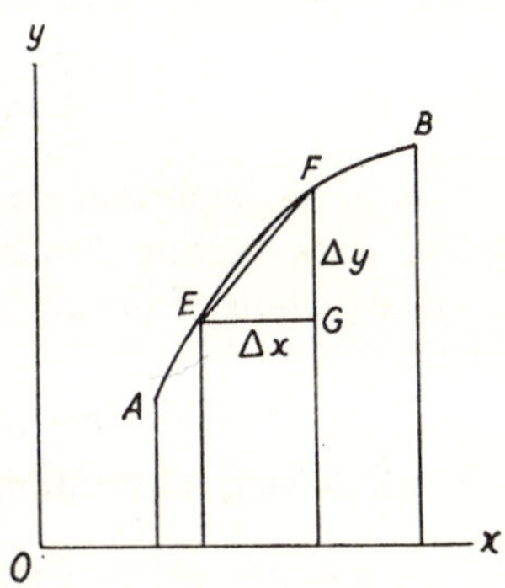

FIG. 48

The triangle EFG is right-angled and therefore

$$(\Delta y)^2 + (\Delta x)^2 = (EF)^2$$

If the point F is moved very close to E then the length of chord EF will be very nearly equal to the length of the arc EF. As F is made indefinitely close to E then, as $\Delta x$ approaches zero, so the length of the chord approaches the length of the arc. Then the equation

$(\Delta y)^2+(\Delta x)^2 = (\text{arc EF})^2$ is approximately true

or $(\Delta y)^2+(\Delta x)^2 = (\Delta s)^2$ ,, ,, ,,

and from this we obtain the equivalent of $\Delta s$ in either of two forms:

$$\Delta s = \sqrt{1+\left(\frac{\Delta y}{\Delta x}\right)^2}\,\Delta x \quad \text{or} \quad \sqrt{1+\left(\frac{\Delta x}{\Delta y}\right)^2}\,\Delta y \quad \text{(approximately)}$$

It may also be shown that

$$ds = \sqrt{1+\left(\frac{dy}{dx}\right)^2}\,dx \quad \text{or} \quad \sqrt{1+\left(\frac{dx}{dy}\right)^2}\,dy \quad \text{(exactly)}$$

This follows because as $\Delta x$ approaches zero so the ratio between chord EF and arc EF approaches 1.

The length of the curve is $s$ and this value may be found by integrating either of the equivalents for $ds$. In practice it will be found that one form usually gives an easier calculation than the other.

Taking the first form of integrand then if $x = a$ and $b$ at points E and F respectively, the length of the arc between those two points is

$$\int_a^b \sqrt{1+\left(\frac{dy}{dx}\right)^2}\,dx$$

This is an important result. Its application may be demonstrated in calculating the circumference of a circle.

The equation of the circle is $x^2+y^2 = a^2$ whence, in the first quadrant,

$$y = \sqrt{a^2-x^2} \quad \text{or} \quad (a^2-x^2)^{\frac{1}{2}}$$

The following substitutions are first made:

$$v = a^2-x^2 \qquad \text{whence } \frac{dv}{dx} = -2x$$

Then $y = v^{\frac{1}{2}}$ whence $\dfrac{dy}{dv} = \frac{1}{2}v^{-\frac{1}{2}} = \frac{1}{2}(a^2-x^2)^{-\frac{1}{2}}$

$$\frac{dy}{dx} = \frac{dy}{dv}\times\frac{dv}{dx} = \tfrac{1}{2}(a^2-x^2)^{-\frac{1}{2}}\times(-2x)$$

$$= \frac{-x}{\sqrt{a^2-x^2}}$$

Therefore $$\left(\frac{dy}{dx}\right)^2 = \frac{x^2}{a^2-x^2}$$

This value may now be used in the formula for the length of an arc. If we again concentrate on the right upper quadrant of the circle then, as in previous examples, the limits within which the expression is to be integrated are $x = 0$ and $x = a$.

The length of the arc which encloses this quadrant is therefore

$$\int_0^a ds = \int_0^a \sqrt{1+\left(\frac{dy}{dx}\right)^2}\, dx$$

$$= \int_0^a \sqrt{1+\frac{x^2}{a^2-x^2}}\, dx = \int_0^a \sqrt{\frac{a^2}{a^2-x^2}}\, dx$$

$$= a\int_0^a \frac{1}{\sqrt{a^2-x^2}}\, dx$$

$$= a\left[\sin^{-1}\frac{x}{a}\right]_0^a \qquad \text{(see Appendix III)}$$

When $x = a$, then $\frac{x}{a} = 1$ and the expression inside the brackets becomes $\sin^{-1} 1$.

But $$\sin^{-1} 1 = \frac{\pi}{2} \qquad \text{(see page 178)}$$

When $x = 0$, then the expression inside the brackets also equals nought. Accordingly the length of arc is

$$a\left(\frac{\pi}{2}-0\right)$$

$$= \frac{\pi a}{2}$$

This is the length of one quarter of the circumference, and the whole circumference is therefore $2\pi a$.

In the above example the working is fairly simple because of the fortunate appearance of an expression which can be simplified to $\sin^{-1}\frac{x}{a}$ and then to $\frac{\pi}{2}$ when $x = a$. The calculation of arcs when such fortunate circumstances do not arise is less simple. However, if the form of the square root may be changed, it may be possible to make use of certain standard integrals. If for example we wish to find the arc length of the curve $y = \frac{x^2}{4}$ within the limits $x = 0$ and $x = 2$ we may proceed as follows:

$$y = \frac{x^2}{4}$$

$$\therefore \frac{dy}{dx} = \frac{x}{2} \quad \text{and} \quad \left(\frac{dy}{dx}\right)^2 = \frac{x^2}{4}$$

Then length of arc is

$$\int_0^2 \sqrt{1+\frac{x^2}{4}}\, dx$$

$$= \int_0^2 \sqrt{\frac{x^2+4}{4}}\, dx \quad \text{or} \quad \frac{1}{2}\int_0^2 \sqrt{x^2+4}\, dx$$

The new expression is a variant of the standard form

$$\int \sqrt{x^2+a^2}\, dx \quad \text{where } a = 2$$

and the integral which we wish to find may be obtained from the standard integral (see Appendix III).

$$\int \sqrt{x^2+a^2}\, dx = \tfrac{1}{2}x\sqrt{x^2+a^2}+\frac{a^2}{2} \log_e (x+\sqrt{x^2+a^2})+c$$

Therefore

$$\frac{1}{2}\int_0^2 \sqrt{x^2+4}\, dx = \frac{1}{2}\left[\frac{1}{2}x\sqrt{x^2+4}+\frac{4}{2} \log_e (x+\sqrt{x^2+4})\right]_0^2$$

Substituting the values of $x = 2$ and $x = 0$ in the normal way and then simplifying and evaluating the logarithm gives the approximate value as 1·8 units of length. These units will of course be measured in the same units as those marked along the $x$ axis.

## SURFACE AREA OF A CONE

Now that we have a method for calculating the lengths of arcs, we may proceed to calculate the areas of curved surfaces of regular solids. This process is best explained first in reference to a circular cone. First we take a wooden cone and label its vertex as A. We then draw a straight line on the surface from the vertex to meet the base of the cone at right angles. This point of intersection is labelled B, and the length of the line drawn is represented by $l$. The cone is now placed on a table so that the line AB is its only line of contact with the table. If the point A on the cone is regarded as remaining at a fixed point on the table and the base of the cone is rolled across the table so that the base completes one full turn, then the base will trace out the sector of a circle as shown in Fig. 49 and the cone will come to rest so that AB now lies along $AB_1$.

In other words, if the surface of the cone could be regarded as a detachable coating then, when removed from the cone, it could be laid out flat in the form of the sector of a circle as shown in the diagram. The calculation of the surface area of the cone is thus resolved into the calculation of the area of a sector of a circle. The radius of this circle is of length $l$, this being the length of AB.

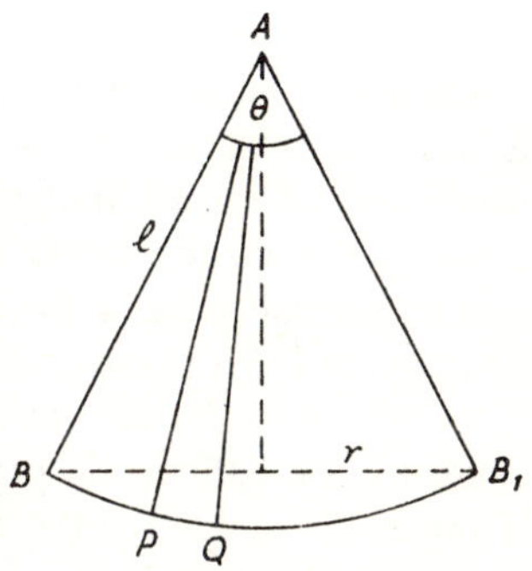

FIG. 49

The arc $BB_1$ is equal in length to the circumference of the base of the cone. If the radius of this base is $r$, then the length of the arc $BB_1$ is $2\pi r$.

We can now calculate the area of the sector described by the rolling of the cone. Let this area be $A$ and the angle $BAB_1$ be $\theta$ radians. Then by the rules of circular measure, the ratio between the area of the sector and the area of the whole circle $\left(\text{that is } \frac{A}{\pi l^2}\right)$ is equal to $\frac{\theta}{2\pi}$ since the whole angle at the centre of the circle is equal to $2\pi$ radians. Thus

$$\frac{A}{\pi l^2} = \frac{\theta}{2\pi}$$

$$\therefore\ A = \tfrac{1}{2} l^2 \theta$$

But again by the rules of circular measure,

$$\theta = \frac{\text{arc}}{\text{radius}} = \frac{2\pi r}{l}$$

Therefore $$A = \tfrac{1}{2}l^2 \times \frac{2\pi r}{l}$$

$$= \pi r l$$

and this gives the formula for the area of the surface of the cone.

The above method does not use calculus at all. The same result is however achieved by calculus as follows. In Fig. 49, PQ is a small part of the circumference of the base of the cone. AS is drawn perpendicular to PQ (not shown in the diagram). Since PQ is a very small part of the circumference it will be very nearly straight. As PQ is made smaller so the sector APQ approaches the form of a triangle, and the area of the sector is approximately equal to one half of the product of the height and base of the triangle. The height of the triangle is AS $= l$, the radius of the circle. The base of the triangle is PQ. The area of the triangle is therefore

$$\tfrac{1}{2}l \times \text{PQ}$$

The area of surface of the cone may be regarded as the limiting value of the sum of the areas of all the small areas like this

$$= \tfrac{1}{2}l \times \text{sum of all arcs like PQ}$$
$$= \tfrac{1}{2}l \times \text{circumference of base}$$
$$= \tfrac{1}{2}l \times 2\pi r = \pi r l$$

## FRUSTUM SURFACE AREA

The next step is to consider the area of the curved surface of a frustum of a cone, the base of which is a circle. A frustum is that part of a cone which remains after the top part of the cone has been lopped off by a cut parallel to the base of the cone. This is represented by the shape AEFB in Fig. 50.

In order to calculate the area of the curved surface of the frustum we regard this area as being equal to the limiting value of the sum of very small trapeziums. This follows logically from the principle involved in treating the surface of the whole cone as consisting of triangles. When the top of the cone is cut off, so also is the top of every triangle cut off leaving a trapezium instead of a triangle.

PQSR is an element of area which approaches the form of a trapezium as PQ is made progressively smaller. The area of a trapezium is equal to half the product of the sum of the parallel sides multiplied by the perpendicular distance between them. Then the area of PQSR is approximately equal to

$$\tfrac{1}{2}a(\mathrm{PQ}+\mathrm{RS})$$

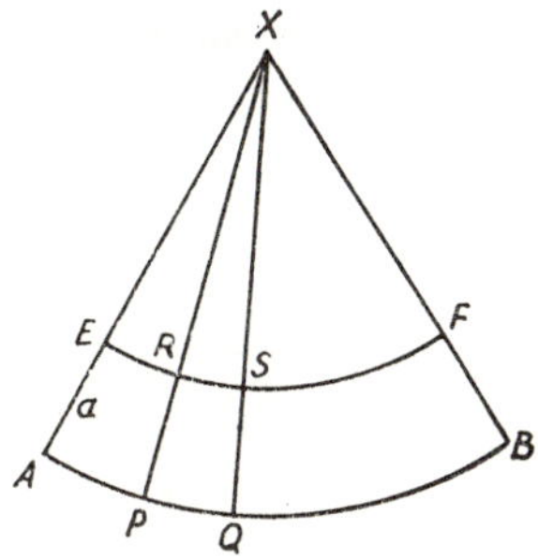

FIG. 50

The area of the frustum surface is equal to the limiting value of the sum of the trapeziums. But, since the base of the cone is a circle, PQ and RS are respectively parts of the circumferences of circles of diameter AB and EF. The surface area of the frustum is therefore equal to

$$\tfrac{1}{2}a(\text{sum of all pieces like PQ}+\text{sum of all pieces like RS})$$
$$=\tfrac{1}{2}a(\text{Circumference AB}+\text{Circumference EF})$$
$$=\tfrac{1}{2}a(2\pi r_1+2\pi r_2) \qquad \text{where } r_1 \text{ and } r_2 \text{ are the respective radii}$$
$$=\pi a(r_1+r_2)$$

or $$2\pi a\left(\frac{r_1+r_2}{2}\right)$$

That is, the area of the curved surface of a frustum is equal to $2\pi$ times the product of the slant side (AE) multiplied by the arithmetic mean of the radii of the circular base and top.

## SURFACE AREA OF A SPHERE

This principle may now be applied to a general consideration of the areas of curved surfaces of all regular solids. In Fig. 51, AB is the diameter of a circle the centre of which is placed at O, the origin of a system of co-ordinates. This diameter thus forms part of the $x$ axis.

PQ is a small part of the curve of the circle. The upper semicircle, when rotated about the $x$ axis, will sweep out the volume of a sphere. The small arc PQ will at the same time generate a part of the surface of the sphere. The points P and Q will describe slightly different circles $C_1$ and $C_2$ as they are rotated. As Q is moved closer to P, so the arc PQ approaches identity with the chord PQ and so the surface then generated by PQ may be regarded as the surface of a cone's frustum of which the circle $C_1$ is the top and circle $C_2$ is the base.

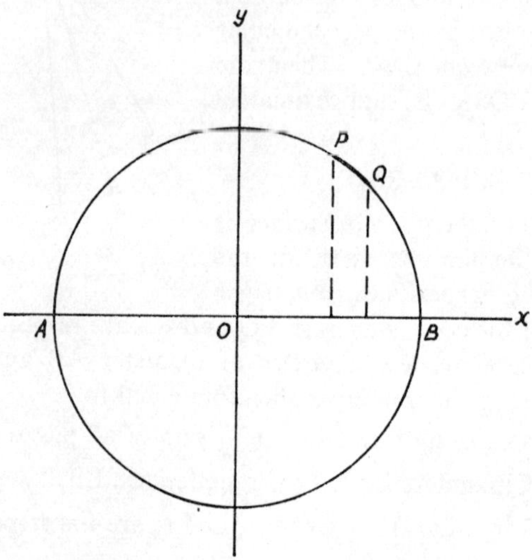


FIG. 51

This being so, the area may be calculated by the formula for the frustum area

$$A = \pi a(r_1+r_2)$$

where $a$ is the length of the slant side and $r_1$ and $r_2$ are the radii of the two circles. In this example the slant side is $\Delta s$ representing an element of surface. The radius of circle $C_1$ is the $y$ ordinate at point P. Thus if this radius is taken as $y$, then the radius of circle $C_2$ will be $y+\Delta y$.

The sum of the two radii is therefore

$$y+y+\Delta y$$

But as PQ is made smaller, so $\Delta y$ approaches zero and the sum of the two radii approaches $2y$.

The surface area generated by an indefinitely small arc PQ is therefore approximately:

$$\pi\,\Delta s(y+y)$$
$$= 2\pi y\,\Delta s$$

the approximation being as close as we like by taking $\Delta s$ sufficiently small. The total surface area generated by the rotation of the whole quadrant is the limiting value of the sum of the areas of all similar strips. That is

$$\text{Area} = 2\pi \int_0^a y\,ds$$

But $$ds = \sqrt{1+\left(\frac{dy}{dx}\right)^2}\,dx$$ (see page 212)

The equation of the curve of the circle, of radius $a$, is (in the first quadrant)

$$y = \sqrt{a^2-x^2}$$

whence $$\frac{dy}{dx} = -\frac{x}{\sqrt{a^2-x^2}} \quad \text{and} \quad \left(\frac{dy}{dx}\right)^2 = \frac{x^2}{a^2-x^2}$$

so $$ds = \sqrt{1+\frac{x^2}{a^2-x^2}}\,dx$$
$$= \sqrt{\frac{a^2}{a^2-x^2}}\,dx$$
$$= \frac{a}{y}dx$$

The formula for the total area may therefore be transformed to

$$2\pi \int_0^a y\left(\frac{a}{y}dx\right)$$
$$= 2\pi \int_0^a a\,dx$$
$$= 2\pi a\Big[x\Big]_0^a = 2\pi a^2-0 = 2\pi a^2$$

This is the area of a hemisphere surface and accordingly the surface area of the whole sphere is twice as great

$$= 4\pi a^2$$

This is exactly four times as great as the area of the original circle, which is the greatest cross-section of the sphere.

In calculating this area we have integrated within known limits and we also finally substituted the particular value of $\frac{dy}{dx}$ in respect of the circle. In its general form however the formula

$$\text{Area} = 2\pi \int y \, ds$$

may be applied to the calculation of the surface area of every regular solid. This formula becomes

$$\text{Area} = 2\pi \int y . \sqrt{1+\left(\frac{dy}{dx}\right)^2} \, dx$$

or

$$2\pi \int y . \sqrt{1+\left(\frac{dx}{dy}\right)^2} \, dy$$

by substituting one of the equivalents of $ds$.

In Chapter 9 it was emphasized that the area under a curve should be regarded as consisting of trapeziums and not of rectangles. The importance of this is now apparent since the trapezium concept is essential to the argument of the theory on which the calculation of surface area is based.

## *Chapter 17*

# INTERPRETATION OF AN AREA

Emphasis has been laid upon the ability of the integral calculus to measure the dimensions of areas of curved plane figures and other shapes. This is of direct use if the areas of such shapes are our main consideration, but very often we are interested in the area under a curve only because of what that area represents.

In the direct case we perhaps need to measure the area of a metal shape in order to calculate how much the shaped metal will weigh. In the indirect case we measure the area under the curve of a function of $x$ because this area represents some other function in which we are interested. The area under a curve may often be accorded a meaning in this way. The nature of the quantity represented by the area will of course depend upon the nature of the variable represented by the curve. This in turn depends upon the variables represented along the $x$ and $y$ axes. We are therefore often in the fortunate position, if we choose our $x$ and $y$ variables correctly, of being able to represent some third variable by the area under the curve.

For example, if we draw the curve of $y = f(x)$ where $x$ represents time in seconds and $y$ represents the velocity of a body in feet per second, then the area under the curve, within the specified limits, will represent the distance travelled by the body. Fig. 52 shows the linear curve of

$$y = 32x$$

This is the same as $v = 32t$, except that the symbols are changed, and in this form may be recognized as the formula for the velocity of a falling stone.

The average velocity of a body is calculated by dividing total distance by total time elapsed. The distance is therefore equal to the product of average velocity and time. The average velocity may be regarded as if it were a constant and the linear curve representing it would be parallel to the $x$ axis. The area under this curve would be in the shape of a rectangle representing $xy$, since the base of the rectangle would be $x$ and the height would be $y$.

In our present example the velocity is not constant. It varies with time. Its curve is therefore not parallel to the $x$ axis and the area under the curve is in the shape of a triangle, as shown in Fig. 52, instead of a rectangle. This area however still represents the distance travelled. It is not equal to $xy$. It is the limit of the sum of the areas of the small strips like $y\,\Delta x$ each of which, for the particular variables represented here, approximately measures the distance travelled within a small interval of time $\Delta x$.

The sum of all these distances is the total distance travelled. This area may be calculated by integration. Since $y = 32x$, then between the limits $x = a$ and $x = b$,

$$\text{Area} = \int_a^b 32x\,dx$$

$$= \Big[16x^2\Big]_a^b$$

In computing areas therefore we may in fact be indirectly evaluating quantities which themselves are not related to areas in the ordinary mensuration sense but which may be represented as if they were. The significant point about this is that, if a quantity may be represented in this way, then it may be evaluated by integration if we know the function of the curve which determines the shape of the area.

In Chapter 9 it was shown that the area under a curve is a function of $x$ such that its derivative provides the equation of the curve. Accordingly, if it is desirable to measure a specified quantity $A$ to be represented by the area, then the curve must have for its equation

$$y = \frac{dA}{dx}$$

In our above example, the curve represents velocity as a function of time. But since velocity is another name for rate of change of distance relative to time, and rate of change is measured by the derivative, this example follows the required pattern. The area $A$ is equal to total distance travelled, and the function of the curve is $y = \frac{dA}{dx}$, the derivative.

This principle may be extended to successive derivatives. Just as velocity is measured by the first derivative of a distance function of time, so acceleration is measured by the second derivative of the same function. But the second derivative is in fact the derivative of the first derivative, and if we draw the curve of the second derivative in respect to time then the area under the curve is equal to the first derivative. That is, if we measure acceleration along the $y$ axis, with time along the $x$ axis, then the area under the curve will represent velocity.

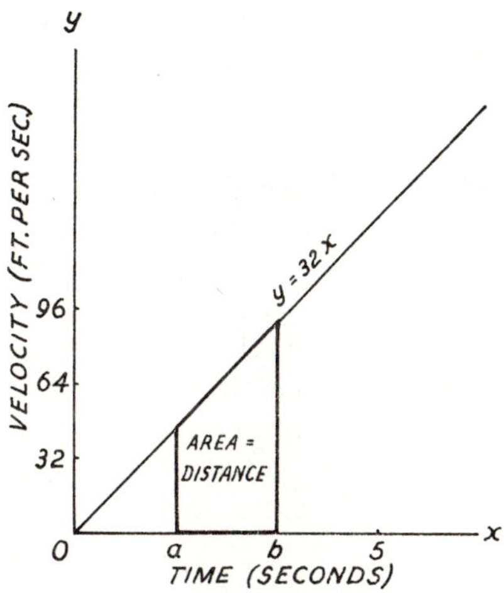


FIG. 52

The acceleration of a falling stone is 32 feet per second. This is constant as it does not depend on $x$. The acceleration curve in this instance is therefore linear and parallel to the $x$ axis drawn through the point $y = 32$. The area under the curve is rectangular in shape and for any value of $x$, the area of the rectangle is equal to $32x$. The speed of a falling stone is also $32x$ (in feet per second).

If this general relationship is viewed from the opposite aspect, then the area under the curve is equal to the integral of the function of the curve. If the function of the curve is $y = x^2$, then the area under the curve is

$$A = \int x^2\,dx$$
$$= \tfrac{1}{3}x^3 + c$$

This shows that the area under the curve can be in the form of a cube number (i.e. when $c = 0$) and suggests that the area under a curve can represent a volume. In simple mensuration it would be impossible to think of an area as if it were a volume since these are quite distinct concepts. In calculus however if we draw the curve of $y = 3x^2$, then the area under this curve

$$A = \int 3x^2\,dx$$
$$= x^3 + c$$

The area under this curve thus represents the volume of a cube of side $x$, when $c = 0$. This representation of course means that the two quantities are numerically equivalent; they have no other similarity.

In general, if a single variable $z$ may be expressed as the product of $x$ and $\bar{y}$, where $x$ is a variable and $\bar{y}$ is the average value over a given range of another variable $y$ (the latter being itself a function of $x$), then $z$ may be represented by the area under the curve of the function $y$ drawn relative to the $x$ and $y$ axes. Then, since $y$ is a function of $x$, so must $z$ be a function of $x$. If we know the value of $x$, we may then calculate the value of $z$.

The area may not always be meaningful in this applied sense. If we draw the curve where $y$ represents units of distance and $x$ represents units of time, then the area represents distance-multiplied-by-time units and this has no meaning which we can interpret since we cannot recognize the so-called units of measurement.

In many instances however the area does have a simple interpretation. An example is provided by the calculation of work done by a variable force. The force is said to do work when its point of application moves in the direction of the force. The unit of measurement of work is the Foot-Pound, and the measure of work done is the product of the force and the distance through which it moves its point of application.

If the force were constant then

$$\text{Work } (W) = \text{Force } (F) \times \text{Distance } (s)$$

When the force varies the equation becomes

$$\text{Work } (W) = \text{Mean Force } (F_1) \times \text{Distance } (s)$$

and, in order to calculate $W$, it would seem that we must first calculate $F_1$ – the mean force. Normally the arithmetic mean is found by dividing the sum of all observed values by the total number of observations. But when the force is varying continuously the theoretical number of possible observations is infinite and such a division is impossible.

The amount of work done may in fact be calculated by integration. If the force is measured along the $y$ axis and distance is measured along the $x$ axis, then the curve of $y$ may be drawn and the area under this curve will represent the total work done within specified limits. That is

$$W = \int F\,ds$$

If a force of 10 lb is required to stretch a spring through 1 inch, what is the work done in stretching it through 3 inches?

By Hooke's Law, the ratio of the required force ($F$) to the distance ($s$) through which the spring is stretched is constant. That is, the longer the stretched spring so the greater is the force required to stretch the spring further.

$$F = K.s$$

When $s = 1$, then $F = K$. But $F = 10$ (when $s = 1$) whence $K = 10$ and $F = 10s$.

The work done is therefore equal to

$$W = \int_0^3 10s\,ds$$

$$= \Big[5s^2\Big]_0^3 = 45 \text{ in-lb} \quad \text{or } 3\tfrac{3}{4} \text{ ft-lb}$$

The energy of the stretched spring may be calculated in exactly the same way since the stored energy is equal to the work done.

The same method may be used to calculate the impulse of a force. Impulse in a given time is equal to the product of the force (if constant) and the time during which it acts. If the force is variable then we can calculate the impulse by measuring the area under the curve of $F = f(t)$. If $F$ = force; $t$ = time; then

$$\text{Impulse} = \int F\,dt$$

In all of the above examples it has been assumed that there is a functional relationship between the variables. If however this is not so then the area under the curve would have to be calculated by Simpson's rule.

It has been shown that although we do not know the value of the mean ordinate of a curve we can still calculate the value of a variable represented by the area under the curve. It is often necessary however to calculate the value of the mean ordinate within specified limits and we may derive this value once we know the size of the area. Thus, instead of using the mean value to calculate the area, we use the area to calculate the mean value.

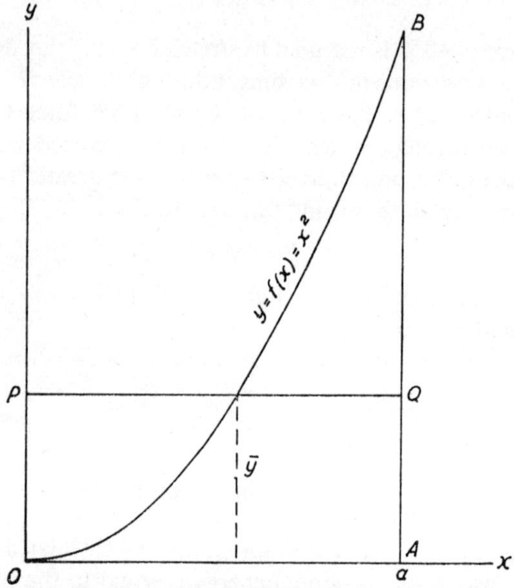


FIG. 53

The mean value is the value of $y$ which, if it had been maintained at a constant level for all values of $x$ within specified limits, would have achieved the same total result as is represented by the area under the curve. The name is perhaps misleading in that the mean value is *not* the arithmetic mean of any finite number of values of evenly spaced ordinates.

Fig. 53 shows the curve of $y = f(x)$. The area between the ordinates $x = 0$ and $x = a$ is

$$A = \int_0^a y \, dx$$

The line PQ is drawn parallel to the $x$ axis so that the area of the rectangle PQAO is equal to the area under the curve. This line cuts the curve at a point where the $y$ ordinate is equal to $\bar{y}$, the mean ordinate. The area of the rectangle is then

$$\bar{y}\times(a-0)$$

But this area is equal to the area under the curve, which is

$$\int_0^a y\,dx$$

Therefore $$\bar{y}a = \int_0^a y\,dx$$

and $$\bar{y} = \frac{1}{a}\int_0^a y\,dx$$

In general, if we wish to calculate the mean ordinate between the limits $x = a$ and $x = b$ (where $b > a$) then the full form is

$$\bar{y} = \frac{1}{b-a}\int_a^b y\,dx$$

The function represented in Fig. 53 is in fact the function $y = x^2$. If $a = 5$, then the $y$ ordinate at this point is 25. By integration the area under the curve is

$$\int_0^5 y\,dx = \int_0^5 x^2\,dx$$
$$= \left[\frac{1}{3}x^3\right]_0^5$$
$$= \tfrac{1}{3}(5^3) = 41\tfrac{2}{3}.$$

The base of the rectangle is 5 units. Accordingly the mean value $\bar{y}$

$$= 41\tfrac{2}{3}\div 5 = 8\tfrac{1}{3}$$

The mean value of a function will not always satisfactorily describe the amount of variation of the function. If the value of the function between the specified limits is negative for some values of $x$ as well as being positive for other values of $x$, then in the calculation of the mean value, these negative and positive values tend to cancel each other out. It was shown in Chapter 15 that the total area under the curve in Fig. 42 (between limits of $x = 0$ and $x = 2$) would be calculated by integration as being equal to zero. This is because part of the curve is below the $x$ axis and the area contained by this part of the curve is negative.

But if the total area is zero, then dividing it by another number to obtain the mean value would give this value as zero also. This of course is meaningless if we are trying to obtain an idea of the variation of the function except in so far as it indicates that some values are positive and some are negative. It is necessary in such instances therefore to calculate some value which does not suffer the disadvantages arising from the cancelling out of negative and positive values.

The *Root-mean-square* or *Quadratic Mean* is just such a value. This value occurs frequently in the study of alternating currents as well as in other branches of physics, and the principle involved has an important application in statistics in the guise of the standard deviation. When describing the voltage of an electricity installation, it is the R.M.S. value which is quoted. Many domestic installations are rated at 240 volts. This means that the R.M.S. value of the supply is 240 volts.

The R.M.S. value is the square root of the mean of the squares of all the values between the specified limits. Each squared value is positive, even if the original value itself is negative, and in summing the squares we are not faced with the predicament of negative and positive values cancelling each other out.

In calculating the mean value we first integrated

$$\int_a^b y\,dx$$

For the R.M.S. value we use the squares of the values and accordingly have to integrate

$$\int_a^b y^2\,dx$$

The mean of this is calculated in the same way as for any other mean value; that is, by dividing by $(b-a)$. Finally, the R.M.S. value is the square root of this mean

$$\text{R.M.S.} = \sqrt{\frac{1}{b-a}\int_a^b y^2\,dx}$$

Another useful ratio in the theory of alternating currents is called the *Form Factor*. This provides a method for describing the form or shape of an alternating current wave. In this connection the word *wave* refers to the wave-like or periodic curve representing the continuous variation of the electro-motive force. This curve is similar to the sine curve shown in Fig. 34 (Chapter 13).

The form factor over a given range is the ratio of the R.M.S. value to the ordinary mean value for the same range:

$$\frac{\text{R.M.S.}}{\text{M.V.}}$$

The ratio is equal to unity for a square wave, this being a curve on which the function alternately assumes two fixed values for equal lengths of time, the time of transition being negligible in comparison with the duration of each fixed value. For a pure sine wave, the ratio is equal to 1·11. The form factor is an important ratio between two quantities which first require to be calculated by integration.

*Chapter 18*

# MECHANICS AND INTEGRATION

The *centre of gravity* of a body is the point through which the line of action of the weight of the body always passes no matter in what position the body is placed. The location of this point is of great practical importance in problems affecting the equilibrium of a body, and the calculus offers a relatively easy way of finding it.

A body may be regarded as consisting of an indefinite number of indefinitely small particles. On each particle is exerted a force due to gravity. This force is proportional to the mass of the particle and is represented by the weight of the latter.

There are then as many forces as there are particles. When two or more forces act upon a body they have a combined effect which may be thought of as a single nett force. This is called the *resultant* of the separate forces. If the forces are parallel and are all acting in the same direction, then the magnitude of the resultant is equal to their sum, and the resultant may be regarded as acting through a point called the *centre of force*.

Since the force of gravity is always acting towards the centre of the earth then, in a small system of particles, the separate forces due to gravity may be regarded as being parallel to each other. The centre of force through which their resultant acts is called the centre of gravity. If $m_1, m_2 \ldots m_n$ represent the masses of the separate particles and $g$ represents the force per unit due to gravity then the separate forces due to gravity will be $gm_1, gm_2 \ldots gm_n$ and their sum is equal to the force of the resultant ($R$). That is:

$$R = gm_1 + gm_2 + \ldots gm_n$$

The points at which the separate forces act may also be related to the centre of gravity in terms of the moments of the respective forces about the same axis. The moment of a force acting on a particle is the product of the force and the perpendicular distance between the particle and the axis. Similarly the moment of the resultant is equal to the product of the resultant force and the distance between its point of application (that is, the centre of gravity) and the axis. This is called the *first moment about the axis.*

If these distances to the $y$ axis are represented by $x_1, x_2 \ldots x_n$ for the particles whose weights are $gm_1, gm_2 \ldots$ etc. respectively, then the respective moments will be $gm_1x_1$, $gm_2x_2$ and so on. Similarly, if the distance between the axis and the centre of gravity is $\bar{x}$, then the moment of the resultant is $R\bar{x}$. But we have seen that $R$ is the sum of the separate forces. Accordingly the moment of the resultant is

$$R\bar{x} = (gm_1+gm_2+ \ldots gm_n)\bar{x}$$

Just as the resultant of parallel forces acting in the same direction is equal to the sum of the separate forces, so also is the moment of the resultant equal to the sum of the moments of the forces. This sum is equal to

$$gm_1x_1+gm_2x_2+ \ldots gm_nx_n$$

There are thus two different expressions for the moment of the resultant and these must be equal to each other. Therefore

$$(gm_1+gm_2+ \ldots gm_n)\bar{x} = gm_1x_1+gm_2x_2+ \ldots gm_nx_n$$

and so

$$\bar{x} = \frac{m_1x_1+m_2x_2+ \ldots m_nx_n}{m_1+m_2+ \ldots m_n}$$

since the $g$ cancels out throughout. That is:

$$\bar{x} = \frac{\text{Sum of moments}}{\text{Sum of masses}}$$

and this expression measures the distance of the centre of gravity from the axis of reference.

If the body is a uniform lamina then its mass may be represented by its area. The sum of the masses is equal to the total mass of the lamina $= \int y\,dx$. Similarly the sum of the moments equals the sum of all the products of each element of area multiplied by its distance from the axis $= \int xy\,dx$. Thus

$$\bar{x} = \int xy\,dx \div \int y\,dx$$

If the axis of reference is the $y$ axis of a co-ordinate system, then the centre of gravity of the body will lie somewhere along the vertical line drawn through the value $\bar{x}$ on the $x$ axis. In order then to obtain a definite fix of the location of the centre of gravity it is also necessary to calculate $\bar{y}$, this being its distance from the $x$ axis. The principles involved in calculating $\bar{y}$ are exactly the same as are involved in calculating $\bar{x}$. The only difference is that we have a different axis of reference. Accordingly

$$\bar{y} = \frac{m_1y_1+m_2y_2+\ldots m_ny_n}{m_1+m_2+\ldots m_n}$$

Since $g$ is eliminated in the final equation, it will be appreciated that the precise location of the centre of gravity does not depend upon the force of gravity. Instead it depends upon the distribution of the particles of a body or, in other words, the distribution of the total mass of the body. Thus, in the equation, mass is related to moments and the latter in turn identify the positions of the particles or, in general, the distribution of all the particles relative to the relevant axes.

Thus $\bar{x}$ and $\bar{y}$ primarily locate the *centre of mass, or centroid.* That they also locate the centre of gravity is secondary to this basic function and further reference here will be to the centroid of a body rather than to its centre of gravity.

The values of $\bar{x}$ and $\bar{y}$ together establish the position of the centroid in relation to the respective axes and it will be noted that each of these values represents the average distance of all particles from the relevant axis. Since we assume that there is an indefinite number of particles, then each of the values to be calculated requires the summation of an indefinite number of indefinitely small values and this may be achieved only by integration.

When setting out to find the centroid of a body which is symmetrical about any axis, it should be noted that the centroid will always lie on this axis. Thus the centroid of a uniformly thin circular lamina coincides with the centre of the circle. The diameter of the circle is an axis of symmetry and the centroid must therefore lie on the diameter. An infinite number of different diameters may be drawn and the centroid must lie on each of them. There is only one point common to all diameters and this is the centre of the circle.

In the same way, if any other body has two intersecting axes of symmetry, then the centroid must coincide with their point of intersection and there is no difficulty in calculating the position of this point relative to the two axes.

Where we wish to locate the centroid of a body having one axis of symmetry then we know that the centroid lies on this axis. It is then necessary only to locate the position of the centroid along this axis. We are already half-way to solving the problems. We have to calculate one distance instead of two.

To simplify the calculations it is usual to take the axis of symmetry as the $x$ co-ordinate axis. Fig. 54 shows a representation of a uniform semi-circular lamina (of radius $a$) so placed that its radius perpendicular to the diameter coincides with the $x$ axis. This radius is an axis of symmetry and the centroid must be at some point on it. This is the point which is a distance of $\bar{x}$ from O, the origin of the axis.

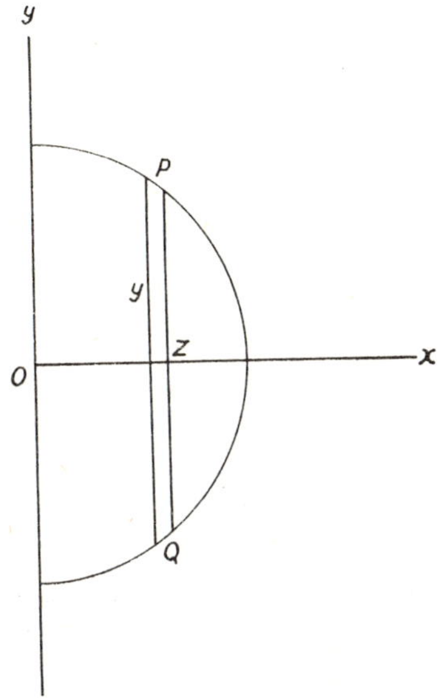


FIG. 54

As the lamina is of uniform thickness its mass may be represented by its area. PQ is a strip of width $\Delta x$ and its area is approximately equal to $2y\,\Delta x$, being twice as great as the strip PZ in the first quadrant. The limiting value of the sum of the areas of all such strips is $\int_0^a 2y\,dx$. This is in fact the area of the semi-circle which we know to be $\frac{\pi a^2}{2}$.

The moment of the strip is equal to its area (as representing its mass) multiplied by its distance from the $y$ axis. That is, the moment $= 2y\,\Delta x.x$. The limiting value of the sum of the moments of all such strips is

$$\int_0^a 2yx\,dx$$



Q

The equation of the circle is

$$x^2+y^2 = a^2$$

whence
$$y = \sqrt{a^2-x^2}$$

(taking the positive root since we have taken the $y$ ordinate in the first quadrant) and the sum of the moments is therefore

$$\int_0^a 2x\sqrt{a^2-x^2}\,dx$$

To integrate this we must resort to the device of substitution.

Let
$$x = a \sin \theta$$

then
$$dx = a \cos \theta \, d\theta$$

We thus have

$$\int_0^a 2x\sqrt{a^2-x^2}\,dx$$

$$= \int_{x=0}^{x=a} 2a \sin \theta \sqrt{a^2-a^2 \sin^2 \theta}\, a \cos \theta \, d\theta$$

$$= 2a^2 \int_{x=0}^{x=a} \sin \theta \, a\sqrt{1-\sin^2 \theta} \cos \theta \, d\theta$$

$$= 2a^3 \int_{x=0}^{x=a} \sin \theta \cos \theta \cos \theta \, d\theta \qquad (\text{since } \cos \theta = \sqrt{1-\sin^2 \theta})$$

$$= 2a^3 \int_{x=0}^{x=a} \cos^2 \theta \sin \theta \, d\theta$$

Another substitution is now required. Let $u = \cos \theta$; then $\frac{du}{d\theta} = -\sin \theta$; and $du = -\sin \theta \, d\theta$.

The integral then becomes

$$2a^3 \int_{x=0}^{x=a} -u^2\, du$$

and this can be integrated by basic principles to

$$2a^3 \left[-\frac{u^3}{3}\right]_{x=0}^{x=a}$$

$$= 2a^3 \left[-\frac{\cos^3 \theta}{3}\right]_{x=0}^{x=a} \qquad (\text{since } u = \cos \theta)$$

$$= -\frac{2}{3}\Big[(a \cos \theta)^3\Big]_{x=0}^{x=a}$$

$$= -\frac{2}{3}\left[\left(a\sqrt{1-\sin^2\theta}\right)^3\right]_{x=0}^{x=a} \qquad (\text{since } \cos\theta = \sqrt{1-\sin^2\theta})$$

$$= -\frac{2}{3}\left[\left(\sqrt{a^2-a^2\sin^2\theta}\right)^3\right]_{x=0}^{x=a}$$

$$= -\frac{2}{3}\left[\left(a^2-x^2\right)^{\frac{3}{2}}\right]_{x=0}^{x=a} \qquad (\text{since } x = a\sin\theta)$$

$$= -\frac{2}{3}(0-a^3)$$

$$= \frac{2}{3}a^3$$

Then, by the principle of moments

$$\bar{x} = \frac{\text{Sum of moments}}{\text{Sum of masses}} = \frac{2}{3}a^3 \div \frac{\pi a^2}{2} = \frac{4a}{3\pi}$$

The centroid of a quadrant of a circle is found in a slightly different way (see Fig. 55). The quadrant does have an axis of symmetry and this is provided by the radius OB which bisects the angle YOX. If this radius were made to coincide with the $x$ axis

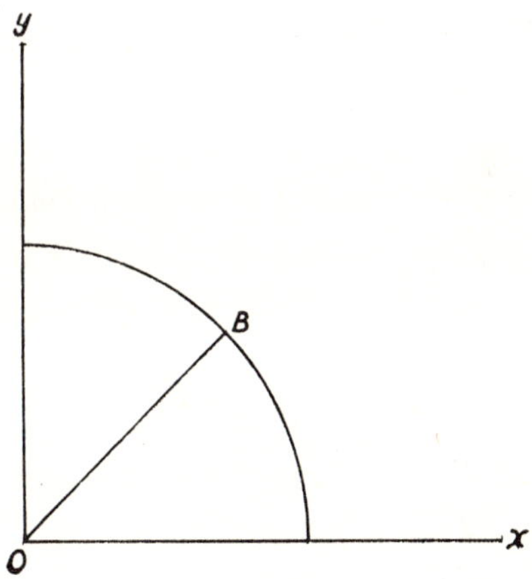


FIG. 55

then we would have a shape above the $x$ axis which would not conform to the curve of a function.

It is to be noted however that since OB is an axis of symmetry then the centroid must be somewhere along this line. Every point on this line is equidistant from each of the co-ordinate axes, so that $\bar{x} = \bar{y}$ and we only need to do one calculation. The sum of the moments

$$= \int_0^a yx\,dx$$

$$= \int_0^a x\sqrt{a^2-x^2}\,dx$$

$$= \left[-\tfrac{1}{3}(a^2-x^2)^{\frac{3}{2}}\right]_0^a = \tfrac{1}{3}a^3$$

This calculation of course follows automatically from the above calculations in respect of the semi-circle.

The sum of the masses is equal to the area of the quadrant and this is equal to one-quarter of the area of the circle. Then

$$\bar{x} = \bar{y} = \frac{\text{Sum of moments}}{\text{Sum of masses}} = \frac{1}{3}a^3 \div \frac{\pi a^2}{4} = \frac{4a}{3\pi}$$

To find the centroid of a solid hemisphere we regard this as a solid of revolution generated by a semi-circle. The centroid lies on the axis of rotation and this is therefore taken as the $x$ axis. In this instance the mass of the body is represented by its volume. The volume of the hemisphere is $\int_0^a \pi y^2\,dx$ which we know is equivalent to $\tfrac{2}{3}\pi a^3$. The sum of the moments of the cylindrical slabs generated by the rotation of the strip PQ (see Fig. 54) is $\int_0^a \pi y^2 x\,dx$. This in turn is equal to

$$= \pi\int_0^a x(a^2-x^2)\,dx = \pi\int_0^a (xa^2-x^3)\,dx$$

$$= \pi\left[\tfrac{1}{2}a^2x^2-\tfrac{1}{4}x^4\right]_0^a$$

$$= \pi(\tfrac{1}{2}a^4-\tfrac{1}{4}a^4) = \tfrac{1}{4}\pi a^4$$

Therefore $\bar{x} = \dfrac{\text{Sum of moments}}{\text{Sum of masses}} = \tfrac{1}{4}\pi a^4 \div \tfrac{2}{3}\pi a^3$

$$= \tfrac{3}{8}a$$

The calculations involved in finding the centroid of the hemisphere are simpler than those involved in finding the centroid of the original semi-circle. This may at first seem strange. The reason of course is that in the example of the semi-circle we have to integrate an irrational expression $\int x\sqrt{a^2-x^2}\,dx$ whereas for the hemisphere we have a much simpler expression $\int x(a^2-x^2)\,dx$ to integrate.

This method of locating the position of the centroid of a body may be used to prove another proposition – that known as *Pappus*' or *Guldin's Theorem*. This concerns the measurement of the volume generated by the rotation of a plane figure about an external axis.

The volume of a straight tube (that is, a cylinder) is equal to the area of its cross-section multiplied by the length of the tube. If we regard a solid ring, such as a curtain ring, as a circular tube, how can the above rule for the measurement of volume be adapted in this special case? We cannot multiply its cross-section by its 'length' since this expression has no meaning unless it is qualified. What is the length of a curtain ring? Its inner circumference is shorter than its outer circumference and neither of these can be taken as the ring's length.

The question then arises as to whether there is some average length which may be calculated and then used to find the volume of the ring. A little thought will suggest that this average might well be provided by the circle which passes through the centre of every cross-section of the ring; that is, a circle which might be regarded as the core of the ring.

The cross-section of the ring is itself circular. We may therefore investigate this possibility by drawing a circle relative to a system of co-ordinate axes and calculating the volume which such a circle would generate if it were rotated about the $x$ axis.

Fig. 56 shows such a circle, of radius $a$, whose centre $c$ is placed in the $y$ axis so that the distance between the centre of the circle and the point O is equal to $b$. We concentrate our attention on the right semi-circle. This meets the $y$ axis at A and at B, and it cuts a line parallel to the $x$ axis at D, where $x = a$. At this point DE is drawn perpendicular to the $x$ axis.

Then the volume generated by the semi-circle is equal to the volume generated by the area ADEO *less* the volume generated by the area BDEO. This method is a simple device in principle but has far-reaching consequences.

In calculating the area ADEO, we regard AD as an arc of the curve of $y_1 = b+\sqrt{a^2-x^2}$. This is explained by the fact that, if

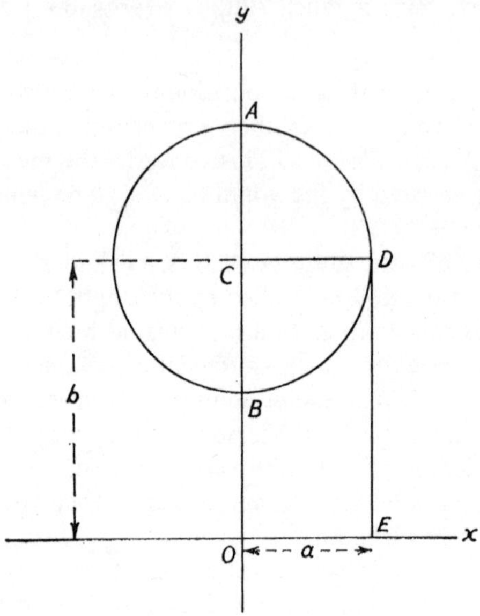


FIG. 56

the circle had been placed so that its diameter coincided with the $x$ axis, then its $y$ ordinates would have been given by $y = \sqrt{a^2-x^2}$. As the circle is moved upwards along the $y$ axis so each of these ordinates increases until, when the circle reaches its position in Fig. 56, each is equal to $b+\sqrt{a^2-x^2}$. The volume generated by ADEO therefore is (by the formula for a volume)

$$V_1 = \pi \int_0^a \left(b+\sqrt{a^2-x^2}\right)^2 dx$$

In the same way we regard BD as an arc of the curve of $y_2 = b-\sqrt{a^2-x^2}$, and the volume generated by BDEO is therefore

$$V_2 = \pi \int_0^a \left(b-\sqrt{a^2-x^2}\right)^2 dx$$

Consequently, the volume generated by the semi-circle is

$$V = V_1 - V_2 = \pi \int_0^a \left(b+\sqrt{a^2-x^2}\right)^2 dx - \left(b-\sqrt{a^2-x^2}\right)^2 dx$$

Multiplying out and simplifying brings this to

$$V = 4b\pi \int_0^a \sqrt{a^2-x^2}\, dx$$

The integration of $\int_0^a \sqrt{a^2-x^2}\, dx$ has already been shown in Chapter 13, giving

$$V = 4b\pi \left[\frac{x}{2}\sqrt{a^2-x^2} + \frac{a^2}{2}\sin^{-1}\frac{x}{a}\right]_0^a$$

$$= 4b\pi \left[0 + \frac{a^2}{2}\cdot\frac{\pi}{2}\right] - 0$$

$\left(\text{This follows because } \sin^{-1}\frac{a}{a} = \sin^{-1} 1 = \frac{\pi}{2}\right)$

Thus $$V = \pi^2 a^2 b$$

The volume generated by the circle – that is the volume of the solid ring – is twice this amount and is equal to

$$2v = 2\pi^2 a^2 b$$

This may be expressed differently as:

$$\pi a^2 . 2\pi b$$

and is therefore equal to the area of the circular cross-section of radius $a$, multiplied by the circumference of a circle of radius $b$. This latter circle represents the path of point C as the original circle is rotated.

The volume of the ring may therefore be calculated by multiplying the area of the cross-section by the length of the circular path described by the centre of the cross-section during one complete rotation about the $x$ axis. The centre of a circle is also its centre of mass, or centroid, since the circle is completely symmetrical. If then we think of the centroid of a rotated plane figure rather than its geometrical centre then this proposition may be extended to all other plane figures whether symmetrical or not.

Thus the volume generated by the rotation of any plane figure about an external axis is equal to the area of the figure (i.e. the cross-section of the resulting solid) whatever its shape, multiplied by the length of the circular path described during one complete rotation by its centroid. This path length may be regarded as the average length of the paths described by all the particles of which the plane figure consists. This of course follows from the fact that the distance of the centroid from the axis is the average distance of all the particles from the axis, and this distance measures the radius of the circular path described on rotation.

That this proposition is true of the solid resulting from the rotation of a regular curve about the $x$ axis may be demonstrated in calculus notation as follows. Let $A$ represent the area under the curve (i.e. the area of the planc figure represented). Then the element of area is $\Delta A$ at approximately $y$ distance from the $x$ axis. The volume swept out by this element will be approximately $2\pi y\,\Delta A$, represented by a narrow tube of length $2\pi y$ and cross-section $\Delta A$. Consequently the total volume swept out by all the elements is the limiting value of their sum:

$$\text{Volume} = 2\pi \int y\,dA$$

But by the principle of moments (where $\bar{y}$ represents the distance of the centroid from the $x$ axis),

$$\bar{y}A = \int y\,dA$$

Therefore, from these two equations,

$$\text{Volume} = 2\pi\bar{y}A$$

and this is equal to the area of the plane figure multiplied by the circumference of a circle of radius $\bar{y}$ which, in turn, is the circle described by the centroid as the figure is rotated about its axis.

The volume generated by the rotation about the $x$ axis of an area under a curve is of course continuous for all values of $y$ within the boundaries represented by the surface of the resulting solid. That is to say, the solid is not in the form of a ring. The principle may however be adapted for a plane figure which does not have the $x$ axis as one of its boundaries (and which will therefore generate a solid ring) by the method employed in the example of the rotation of the circle. But instead of using $b$ as the distance between the $x$ axis and the circle's centre, we use $\bar{y}$ as the distance between the $x$ axis and the centroid of the rotated figure.

This proposition was first made by the Greek mathematician Pappus and much later was again made by Guldin. Guldin is alleged by some to have adopted Pappus' theorem as his own although, on his behalf, it has been claimed that Guldin's theorem could not have been deduced from the form in which it was given by Pappus.

This is a very minor mathematical disputation and, for practical purposes, we are not really concerned whether Pappus or Guldin should be credited with the theorem. Some doubt has even been expressed as to whether Pappus actually originated his theorem or whether he was merely reporting another's discovery! But these few comments may help to explain that the theorems bearing the name Pappus and of Guldin are in fact one and the same.

## MOMENT OF INERTIA

Another very important unit in mechanics is the Moment of Inertia of a body rotating about an axis. A good example of such a body is provided by the flywheel which, because of its turning power when in motion, stores up kinetic energy which it later releases to equalize the motion of an engine.

Inertia of matter is the property by which it offers resistance to a change in its state of rest or in its state or direction of motion. If we wish to make a body rotate about some axis then a force must be applied to it. If the force is applied close to the axis it will not have as great a rotary effect as would an equal force applied at a point further away from the axis.

The effect of the body in rotation about the axis depends upon the way in which the mass is distributed as well as upon the total amount of mass in motion. This combined effect is measured as the Moment of Inertia of the body about the axis and is usually represented by the symbol $I$.

The moment of inertia of a small particle of mass $m$ is mathematically defined as

$$I = mr^2$$

where $r$ represents the perpendicular distance between the mass and the axis. As this moment involves the *square* of $r$ it is called the *second* moment about the axis. It must not be confused with the *first* moment which is used in the calculations to locate the centroid of a body.

If we regard the body as consisting of a number of very small parts each of mass $\Delta m$, then the moment of inertia of each part is approximately $r^2 \Delta m$. The total moment of the body is the limiting value of the sum of the moments of the component parts. That is:

$$I = \int r^2 \, dm$$

This quantity enables us to calculate the kinetic energy of the rotating body.

Kinetic energy is the power of doing work possessed by a body in virtue of its motion. If a driver releases the clutch of a car moving at speed, so that the engine no longer has any direct effect on the forward motion of the car, the car will nevertheless continue to move forward by virtue of its speed. The mass contained in the structure of the car will have stored up energy because of its motion and, on a level road surface, the car will not stop of its own accord until that energy has been consumed in the continued unassisted forward motion.

If a body of mass $m$ is moving in a straight line with velocity $v$, then its kinetic energy is equal to $\frac{1}{2}mv^2$. The kinetic energy of a very small piece is therefore approximately $\frac{1}{2}v^2 \Delta m$. But if a body is rotating about an axis, which itself is not moving forward, then it apparently has no forward velocity and its kinetic energy cannot be calculated from the above formula.

As in other calculus problems therefore it is necessary to concentrate our attention on a particle of mass. If a body of mass $m$ is rotating with an angular velocity (i.e. in radians per second) of $\omega$ about an axis then, at any given instant, a particle $\Delta m$ is moving with a linear velocity $v$, where $v = r\omega$ ($r$ being the distance of the particle from the axis).

The kinetic energy of the small piece is approximately $\frac{1}{2}v^2 \Delta m$. Since $v = r\omega$, then the energy is equal to $\frac{1}{2}(r\omega)^2 \Delta m$. The kinetic energy of the whole body is equal to the limiting value of the sum of the energy of all such pieces. Thus

$$\text{K.E.} = \int \tfrac{1}{2}(r\omega)^2 \, dm$$

$$= \tfrac{1}{2}\omega^2 \int r^2 \, dm$$

But $\int r^2\,dm$ measures the moment of inertia of the body and accordingly

$$\text{K.E.} = \tfrac{1}{2}\omega^2 \times I$$

The necessary calculations are now shown in a simple example. Fig. 57 shows a thin uniform rod of length $a$ (= AB) and density $\rho$ placed so that its longitudinal axis of symmetry coincides with the $x$ axis. If the rod rotates with angular velocity equal to $\omega$ about

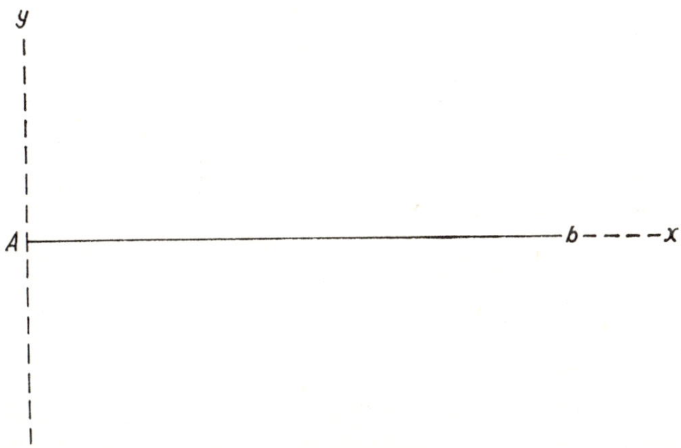


FIG. 57

the axis at A, what is its moment of inertia and what amount of work can it do in being brought to rest? The mass of a very small piece of the rod of length $\Delta x$ is equal to $\rho\,\Delta x$. Its distance from the axis of rotation is $x$. Its moment of inertia is approximately $(\rho\,\Delta x)(x^2) = \rho x^2\,\Delta x$. The total moment of inertia is then the limiting value

$$\int_0^a \rho x^2\,dx = \rho\left[\frac{x^3}{3}\right]_0^a = \frac{\rho a^3}{3}$$

The work it can do in being brought to rest (that is, its kinetic energy) is

$$\tfrac{1}{2}\omega^2 \times I = \tfrac{1}{2}\omega^2 \times \frac{\rho a^3}{3}$$

$$= \frac{\rho a^3 \omega^2}{6}$$

If a circular lamina of uniform density is rotating about an axis through its centre perpendicular to the plane of the lamina, what is the moment of inertia? Let the mass of the lamina be $m$ and the radius of the circle be $r$. This is shown in Fig. 58, it being understood that the lamina rotates about the $y$ axis at O. The shaded band of width $\Delta x$ represents the element of area placed at a distance $x$ from the axis of rotation. Since the lamina is of uniform density its mass may be represented by its area. This of

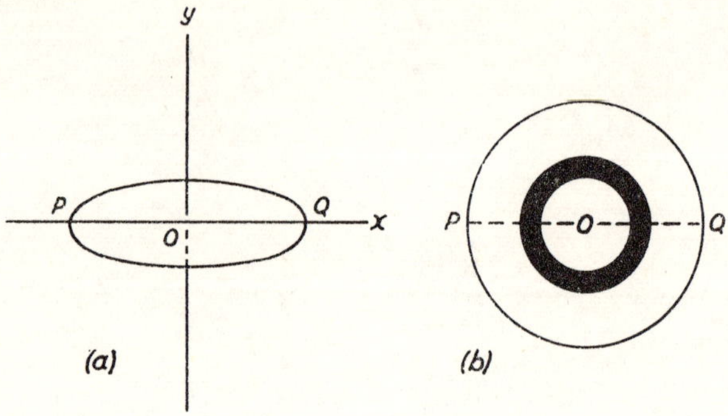


($a$) side view of lamina ($b$) lamina viewed from the top

FIG. 58

course also applies to the shaded band. The moment of inertia of this band is approximately $2\pi x \, \Delta x \times x^2$. The moment of inertia of the total area is the limiting value of the sum of the moments of all bands like this.

$$I = \int_0^r 2\pi x^3 \, dx = 2\pi \left[ \tfrac{1}{4}x^4 \right]_0^r$$
$$= \tfrac{1}{2}\pi r^4$$

But the total mass ($m$) represented by the area $= \pi r^2$. Therefore

$$I = \tfrac{1}{2}mr^2.$$

## PRESSURE

Another useful application of integral calculus arises in the calculation of the total force exerted on a body by liquid with which it is in contact.

Fig. 59 shows a rectangular plate of size $b \times c$. It is immersed in a tank of liquid of density $\rho$ and is placed so that it remains vertical and so that its top edge just cuts the surface of the liquid. What is the total force exerted on one side of the plate by the liquid?

For reasons which will become apparent, we first consider the downward pressure of the liquid. This pressure varies with the depth of the liquid. It increases as the depth is increased. That is, the pressure at a point is directly proportional to the depth of the point below the surface of the liquid. This of course explains why

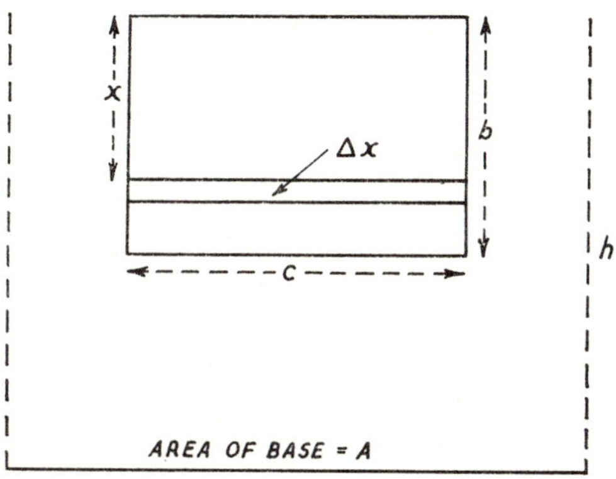


FIG. 59

dock gates, reservoirs, and canal banks are made stronger towards the base of their constructions so as to withstand the greater pressure.

The downward force on any horizontal plane, such as the base of the tank, is equal to the weight of a vertical column of liquid the cross-section of which is equal to the area of the base. If the depth of the liquid in the tank is $h$ and the area of the base is $A$, then the total force exerted by the liquid on the base is $Ah\rho g$ (area of base $\times$ depth of liquid $\times$ density of liquid $\times$ unit force of gravity).

The pressure is identical at every point in the base and the above expression represents the sum of the pressures at all points. This of course is not true of the vertical rectangle. The pressure varies as between points at different depths in the liquid.

One important property of liquids is that at any point in the liquid the pressure is the same in all directions. Thus at any point at depth $x$, the sideway and downward pressures are the same. We therefore consider a small horizontal strip of area at depth $x$ representing part of the plate. If the vertical depth of this strip be $\Delta x$ and its width is $c$, then its area is $c\,\Delta x$. At depth $x$ the downward force on an area of this size would be approximately $\rho g \times x \times c\,\Delta x = \rho gcx\,\Delta x$. But the sideway pressure at point $x$ is the same as the downward pressure, and the total force is equal to the limiting value of the sum of the pressures on all the strips of area at points between the limits $x = 0$ and $x = b$.

The required force is therefore equal to

$$\int_0^b \rho gcx\,dx = \rho gc\left[\frac{x^2}{2}\right]_0^b = \frac{\rho gcb^2}{2}$$

## CENTRE OF PRESSURE

Just as the centre of gravity of a body is the point of action of the resultant of all forces acting on the body due to gravity, so the *centre of pressure* is the point on a surface in contact with a liquid at which the resultant fluid pressure acts. The position of the centre of pressure is important in matters of ship design and stability. As a craft makes its way in water there will be other forces acting on its keel. Its design must provide for the balancing of forces and, in order to be able to do this, the designer must know at what point the pressure due to the water is acting.

The centre of pressure on the plate in the above example may be found as follows. The elements of area considered are all parallel and the sideway pressure forces acting on them are parallel. The position of the centre of pressure may therefore be calculated by using the same formula as for the centre of gravity. This is because the forces in each instance are parallel forces, and the formula applies to parallel forces generally. The centre of gravity and the centre of pressure are just special kinds of centres of force.

Let $\bar{x}$ be the depth at which the centre of pressure acts. The moment of the force acting on each element of area is equal to

$$\text{force} \times \text{depth}$$

$$= \rho gcx\,\Delta x \times x = \rho gcx^2\,\Delta x \quad \text{(approx.)}$$

Then the limiting value of the sum of all the moments is

$$\int_0^b \rho gcx^2\,dx = \rho gc\left[\frac{x^3}{3}\right]_0^b = \frac{\rho gcb^3}{3}$$

and
$$\bar{x} = \frac{\text{Sum of moments}}{\text{Total pressure}}$$

$$\therefore\ \bar{x} = \int_0^b \rho gcx^2\,dx \div \int_0^b \rho gcx\,dx$$

$$= \frac{\rho gcb^3}{3} \div \frac{\rho gcb^2}{2}$$

$$= \frac{2b}{3}$$

The centre of pressure is therefore two-thirds of the way down from the top of the plate.

## CYCLOID

The cycloid is the curve traced out by a fixed point on the circumference of a circle which rolls without slipping in contact with a straight line. In practical applications it may be thought of as being generated by a point on the rim of a wheel.

In Fig. 60, P is a point on the circumference of the circle (centre C) which is rolled along line DBA. The movement of point P generates the cycloid of which DOA forms one complete section. The line OB is taken as the $y$ axis and a straight line perpendicular to it through O is taken as the $x$ axis.

T is the variable point of contact of the circle with the $x$ axis. Join PC and CT to form the angle PCT $= \theta$. This angle varies as the circle is rolled along. CT is always perpendicular to the $x$ axis, whereas PC rotates as the circle is rolled.

The equations of the cycloid may be expressed in terms of $\theta$ as:

$$x = a(\theta+\sin\theta)$$
$$y = a(1-\cos\theta)$$

where $a$ = radius of circle.

The area of one section of the cycloid may be calculated as follows (see Fig. 61). Draw the perpendicular FPG. This is equal to the diameter of the circle so that, if the radius of the latter is $a$ then FPG $= 2a$. But PG $= y$. Therefore FP $= (2a-y)$.

Next, BA is equal to one half of the circumference of the circle and the value of $x$ at A is therefore $\pi a$. The integration of the area of the right half of the cycloid must therefore be effected between

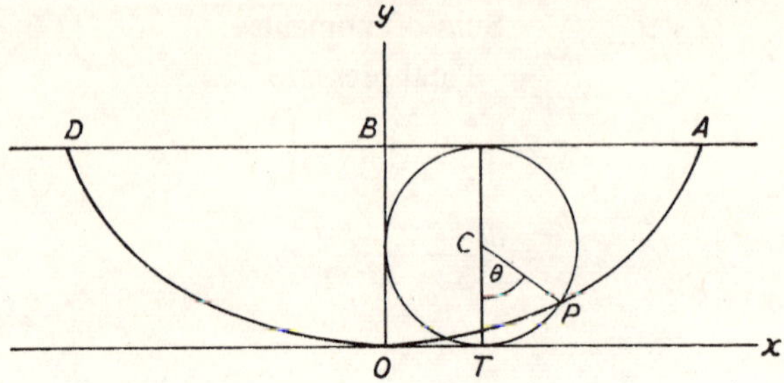


FIG. 60

limits of $x = 0$ and $x = \pi a$. An element of area of this half of the cycloid is equal to (FP)($\Delta x$). But FP $= 2a - y$. These elements of area are now summed and, finally, the limiting value of this

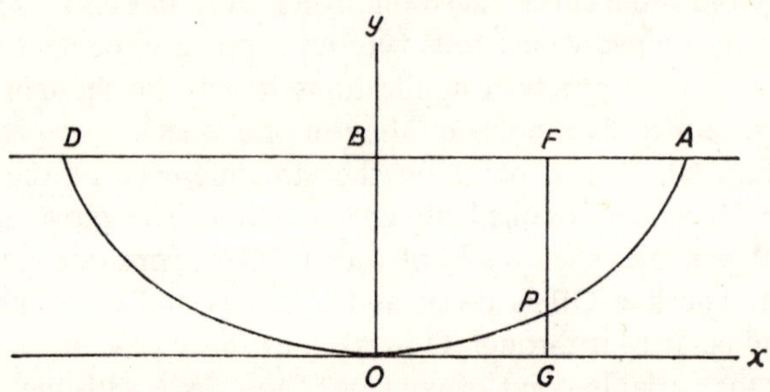


FIG. 61

sum is doubled to give the area of the whole section DOA. Thus

$$\text{Area} = 2\int_0^{\pi a} (2a-y)\, dx$$

But $$y = a(1-\cos\theta)$$

so $$(2a-y) = 2a - a(1-\cos\theta)$$
$$= a(1+\cos\theta)$$

and $$x = a(\theta+\sin\theta)$$

so $$dx = (a + a\cos\theta)\, d\theta$$
$$= a(1+\cos\theta)\, d\theta$$

Substituting these values for $(2a-y)$ and $dx$, the area may be expressed in terms of $\theta$. The limits must also be changed. When $x = \pi a$; $\theta = \pi$. When $x = 0$; $\theta = 0$. The limits, in terms of $\theta$, are therefore $\pi$ and 0.

The area may now be expressed

$$2\int_0^\pi a(1+\cos\theta)a(1+\cos\theta)\,d\theta$$

$$= 2a^2\int_0^\pi (1+\cos\theta)^2\,d\theta$$

$$= 2a^2\int_0^\pi (1+2\cos\theta+\cos^2\theta)\,d\theta$$

But $\qquad \cos^2\theta = \frac{1}{2}(1+\cos 2\theta) \qquad$ (see Appendix II)

$$\therefore \text{Area} = 2a^2\int_0^\pi \left(1+2\cos\theta+\frac{1}{2}+\frac{1}{2}\cos 2\theta\right)d\theta$$

$$= 2a^2\int_0^\pi \left(\frac{3}{2}+2\cos\theta+\frac{1}{2}\cos 2\theta\right)d\theta$$

$$= 2a^2\left[\frac{3}{2}\theta+2\sin\theta+\frac{1}{4}\sin 2\theta\right]_0^\pi$$

When $\theta = \pi$, $\sin\theta = 0$; and $\sin 2\theta = 0$. Then

$$\text{Area} = 2a^2\left(\frac{3}{2}\pi+0+0\right)-0 = 3\pi a^2$$

Thus the area is exactly three times as great as the area of the generating circle. It is interesting to note that, although it is impossible to calculate the exact area of a circle – since any calculation involving $\pi$ is necessarily approximate – we can nevertheless construct an area which is *exactly* equal to the area of the circle. For example, let the circle in Fig. 60 be placed so that its centre C is on the $y$ axis, then half of the area falls within BAO. But BAO is equal to one-and-a-half times the area of the circle. The difference between the area of BAO and the area of the semi-circle is therefore exactly equal to the area of the circle and is represented by the enclosed space between the boundaries provided by the circumference of the circle, the $x$ axis and the cycloid.

The cycloid occurs in a number of applications. One of its very useful properties arises from the fact that the arc of a cycloid provides the path of quickest descent between two points not in the same horizontal plane. If there are two fixed points A and B, then a sphere which is allowed to roll along a path from A to B will move most quickly if its path is in the shape of an inverted cycloid. Its progress will be even more rapid than if it had rolled along a flat inclined surface despite the fact that its cycloid path is necessarily longer.

## *Chapter 19*

# STATISTICS

Most statistical techniques and measurements depend upon or are related to the properties of curves and require the interpretation and measurement of the areas under those curves. It is not surprising therefore that calculus should play its part in helping with the calculations as well as assisting in the development of the theory.

The underlying theories are in general too complex to be discussed here, but it is proposed to show how the area under a curve may be interpreted in terms of probability, how this and other measures may be calculated and how one of the theories may be used in a specific application.

Probability is itself a separate branch of mathematics. Here we restrict our interpretation to the view of Aristotle who said that the probable was that which usually happened. In line with this view we say that the relative frequency with which an event has been observed also measures the probability of that event continuing to occur, provided the conditions in which it has occurred remain the same.

If there are two possible outcomes of an event and one outcome has occurred twice as frequently as the other over a long period then, in the absence of any known change in the conditions affecting the event, we would still expect the same ratio between the frequencies of the two outcomes in the future.

In order to discuss the probability of an outcome we must therefore have details of the frequencies. If we measure the height of every soldier in a group of 50,000 we will find that most of the measurements cluster round the value of the arithmetic mean of all the measurements. Furthermore the frequency of measurements tails off in a fairly rapid manner as the differences between the mean and the individual measurements increase.

These measurements could be shown in the form of a histogram as in Fig. 62. Height is measured along the $x$ axis in class intervals and the frequency (or number of measurements recorded in each class interval) is measured along the $y$ axis. A frequency polygon is constructed by joining together the mid-points of the tops of the histogram bars.

Histograms are very useful as pictorial representations of facts, but they do not lend themselves to mathematical processes in the same way as do the smooth curves of functions. The statistician

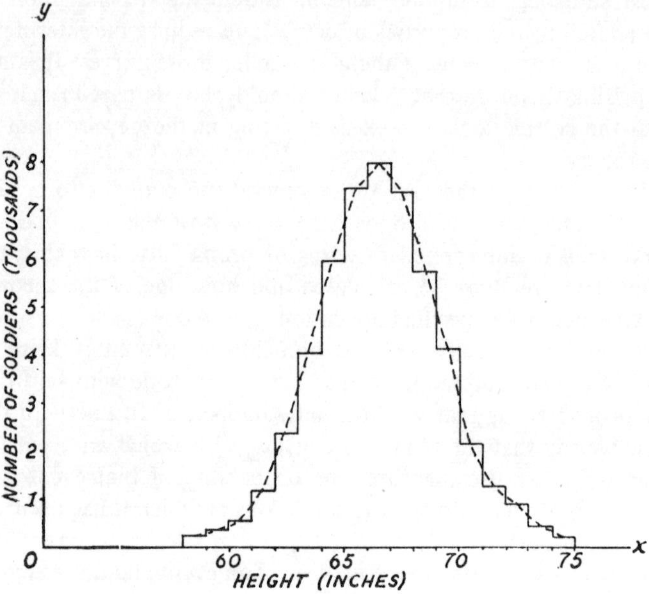


FIG. 62

therefore calls into use the concepts of geometrical approximation and limiting values. He admits that his histogram is not a curve but he claims to be able in many instances to identify a curve which will satisfactorily replace the histogram. How justifiable is this claim?

The difference between discrete and continuous variables has already been noted in Chapter 2. Height is a continuous variable in that it can take *any* value within the range of values, but because of the difficulty of measuring it *exactly* it is necessary to show the measurements in class intervals. The results therefore appear to show height as a discrete variable in that the class limits 'jump' from one value to another. The continuity of the variable however exists independently of our ability to measure it and we are therefore justified in treating height as a continuous variable.

We may now therefore conceive of a very small class interval which would be represented along the $x$ axis by $\Delta x$. If we could draw the histogram using a class interval of this size the bars would be very narrow and the frequency polygon would be smoothed out appreciably. If the histogram bars were drawn progressively narrower (that is, as $\Delta x$ is made smaller), the frequency polygon would approach ever more closely to the form of a smooth curve. The tops of the bars would indeed be indistinguishable from their own mid-points and the area under the curve would be equal to the limiting value of the sum of the histogram bars.

Such a curve would be something like the curve shown in Fig. 63 and would represent the frequency distribution of the measurements, in that it would show how the individual measurements are distributed over the range of measurement.

This is more satisfactory than the histogram since it more specifically provides for potential measurements, that is to say the values which cannot be measured exactly. We cannot pretend however that the curve exactly describes a mathematical function, as we are in effect charting frequency as a function of height. Height is a continuous variable but frequency measurements are discrete. Frequency can be expressed only in whole numbers. We cannot, for example, say that the height of 5 feet 4 inches occurs $1\frac{1}{4}$ times. Nevertheless we may treat this relationship between height and frequency as being so nearly equivalent to a functional one as will make no practical difference in our results. When we are dealing with 50,000 measurements, the error in regarding a value of $y = 1\frac{1}{4}$ as indicating a frequency $= 1$ is negligible.

At this stage it is appropriate to mention that statistics do not claim to give exact answers to any question, although they do seek to give very good approximations in problems for which exact quantities cannot be calculated. The curve itself is only an approximate equivalent of the underlying histogram and it is of course an idealized approximation.

Because of the facility for approximating the area under a curve to the area of a histogram, it is common to treat discrete variables as if they are continuous. Continuity requires the imperceptible merging of one value with the next value. In order that a discrete variable may be treated as continuous therefore it is necessary that its total frequency should be large and that the intervals of its measurements should be very small. When the size of the intervals is extremely small relative to the number of intervals and the total frequency the desired identity between discrete and continuous variables is approximately realized.

Here again then it is not claimed that this principle can give exact results, but it often gives results which approximate sufficiently closely to results obtained by the much more complex methods in the techniques required if continuity is not assumed.

The curve in Fig. 63 is in fact a representation of the *Normal Curve* and the distribution which it describes is therefore called a *normal distribution*. This is one of the most important curves used in statistics. It was first derived by de Moivre but is more usually associated with the name of Gauss in connection with his Normal law of errors which states that chance errors in measurements tend to be normally distributed. It has been found by experiments that many kinds of scientific and sociological data approximate fairly closely to the normal, but it should be noted that there is nothing necessarily abnormal about a distribution which does not 'fit' the normal curve – it may merely belong to a different family of distributions.

The curve is symmetrical about the $y$ ordinate drawn at the point on the $x$ axis which corresponds to the value of the arithmetic mean of all the measurements. It will be recalled that the centroid of a lamina is necessarily situated on an axis of symmetry. Accordingly the centroid of a lamina in the shape of the normal curve must lie in the $y$ ordinate drawn at the value of $x$ equal to the arithmetic mean. The distance of the centroid from the axis is usually represented as $\bar{x}$, this representing the average distance from that axis of all the particles of which the lamina is composed. The same symbol $\bar{x}$ is carried over into statistical calculations where it represents the arithmetic mean of the distribution.

The equation for locating the $y$ ordinate in which the centroid is situated is

$$\bar{x} = \int xy\,dx \div \int y\,dx$$

and accordingly the same equation may be used to calculate the arithmetic mean of a continuous distribution.

The equation of the normal curve is

$$y = \frac{1}{\sigma\sqrt{2\pi}} e^{\frac{-(x-\bar{x})^2}{2\sigma^2}}$$

where $\sigma$ is the standard deviation.

The difference between an individual measurement and the arithmetic mean of the distribution is called the deviation of that measurement. The standard deviation of the whole distribution is calculated as the root-mean-square of all the deviations and thus measures the dispersion of all the measurements about the mean.

The mean of all the squared deviations is called the variance. This, by the above definition, is equal to the square of the standard deviation and is represented by $\sigma^2$. The variance is calculated as the second moment of the distribution about the $y$ axis and is therefore analogous to the moment of inertia in mechanics. Each of these quantities describes the dispersion of values about the relevant axis.

The equation for the variance is

$$\sigma^2 = \int (x-\bar{x})^2 y \, dx \div \int y \, dx$$

By expanding the binomial $(x-\bar{x})^2$ it is possible to transform this equation into a form which will make the actual calculations simpler.

$$\int (x-\bar{x})^2 y \, dx = \int (yx^2 \, dx - 2x\bar{x}y \, dx + \bar{x}^2 y \, dx)$$

$$= \int yx^2 \, dx - 2\bar{x} \int xy \, dx + \bar{x}^2 \int y \, dx \qquad \text{(i)}$$

But $$\bar{x} = \int xy \, dx \div \int y \, dx$$

so $$\int xy \, dx = \bar{x} \int y \, dx$$

and the right-hand side of equation (i) becomes

$$\int yx^2\,dx - 2\bar{x}^2\int y\,dx + \bar{x}^2\int y\,dx$$

$$= \int yx^2\,dx - \bar{x}^2\int y\,dx$$

$$\therefore \sigma^2 = \left(\int yx^2\,dx - \bar{x}^2\int y\,dx\right) \div \int y\,dx$$

$$= \frac{\int yx^2\,dx}{\int y\,dx} - \bar{x}^2$$

The standard deviation is the square root of this expression. In the above transformation the symbol $\bar{x}$ was first moved to the left of the integral sign. This is because although $x$ varies, $\bar{x}$ is a constant.

The usefulness of the curve may now be shown in an application to the quality control of a process manufacturing small metal rods. The process is designed to produce rods of a specific length; suppose this to be 1 inch. No process is perfect however and this process turns out rods which vary within very small limits from the desired measurement. If the process is continuous in unchanging circumstances then the population (i.e. total number) of rods may be regarded as infinite. Length is a continuous variable and we may apply the properties of some curve to its distribution.

Over a long period it is found that the mean length of all rods produced is in fact 1 inch and that the standard deviation is equal to ·001 inches. In theory therefore all future production would be expected to conform to the same distribution unless something goes wrong in the process. The object of quality control in this instance is to set up a warning system so that any fault which arises in the process may be detected as soon as possible.

This is where the properties of the normal curve prove their usefulness. For any given number of measurements the area under the curve $\left(\text{area} = \int y\,dx\right)$ measures the total frequencies of all the measurements taken together. Thus every measurement is represented within that area. We now regard the relative frequency of a measurement as its probability ratio. As the curve is symmetrical then 50% of all measurements will occur in the right-hand half of the area, and 50% will occur in the other half. Then the probability of a recorded measurement being in, say, the right-hand half is said to be equal to 1 chance in 2. This probability is expressed in the ratio $\frac{1}{2}$.

Similarly the probability that a measurement will occur within any two $y$ ordinates is represented by the area of the curve between those ordinates expressed as a proportion of the total area. The probability of a measurement falling between the values $x = a$ and $x = b$ is therefore

$$P = \int_a^b y\,dx \div \int_{-\infty}^{\infty} y\,dx$$

The total area 'under' the curve is shown as to be integrated between the infinite limits of $\infty$ and $-\infty$ because the curve never actually meets the $x$ axis. This might appear to introduce some complications in the actual integration but in fact we can side-step these. We know by definition that the probability of a measurement falling within the total area must be equal to 1 chance in 1. The probability ratio for this is therefore equal to $\frac{1}{1} = 1$. Accordingly the total area may also be taken as being equal to 1 since this area represents the total probability.*

Thus

$$\int_{-\infty}^{\infty} y\,dx = 1$$

Consequently the probability of a measurement falling within the quoted limits may now be simplified to

$$P = \int_a^b y\,dx$$

The problem is thus converted to one of integrating the function of the curve between specified limits. Since the function $y$ is in fact a complicated one it might be thought that converting the problem to one of integration would make it a more difficult rather than an easier one. And so it would if we had to perform the integration each time for every problem. Fortunately there is no need

* See C. E. Weatherburn, *First Course in Mathematical Statistics* (Cambridge U.P., 1961), p. 51.

to do this. It has already been done for us and we can find the answer by looking in the appropriate set of mathematical tables which will be found in any textbook on statistics.

There is not just one normal curve and it is wrong to refer to '*the* normal curve' in general terms as if it were a specific and unique curve. The description refers to a type of curve rather than to an individual curve. Nevertheless there is a standard normal curve for which mathematical dimensions have already

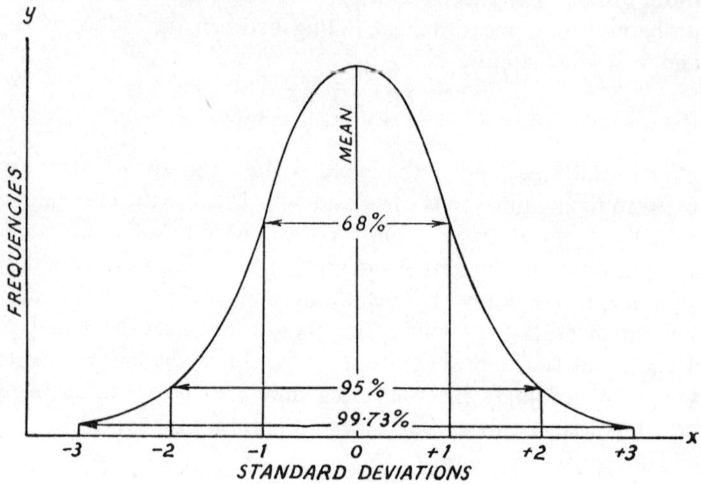


FIG. 63

been calculated. Other normal curves may be related to the standard to derive their own particular dimensions.

Fig. 63 shows a notional distribution for the metal rod measurements. The $x$ axis has been marked off in units of the standard deviation from the mean. The reason for this is that it may be shown that about 68% of the total area is contained within ordinates drawn on either side of the mean at 1 standard deviation distance from the mean. The probability that a measurement will lie within the same limits is therefore also equal to 68 chances out of 100. The probability ratio is thus $\frac{68}{100} = 0{\cdot}68$.

In the same way about 95% of the area lies within $\pm 2$ standard deviations of the mean, and about 99·73% lies within $\pm 3$ standard deviations. These properties are shown in Fig. 63.

This information may now be applied to the quality control requirements. Since 99·73% of all the measurements lie within ±3 standard deviations, then less than 3 out of every 100 measurements would be expected to fall outside these limits. If all the observed measurements do in fact lie within the control limits set up at ±3 s.d. from the mean, then the process is said to be in control. If measurements occur outside the limits then, because of the very low probability of this happening, we have an indication that something may have changed in the process. If a further measurement outside the limits is recorded then there is a positive indication that something has changed in the process itself. Appropriate action may be taken promptly to remedy any fault which may be identified as a result of the statistical warning.

In practice it is usual to take small samples and to chart the arithmetic means of the samples rather than the measurements of the individual items inspected. There are very good reasons for this but they do not concern us here. The sample means themselves form a separate distribution, called the *sampling distribution*, with its own mean, standard deviation and control limits.

Actual distributions are rarely, if ever, exactly symmetrical, so that the normal curve represents an idealized form of distribution. Nevertheless many distributions conform very closely to the normal and may be regarded as possessing normal characteristics. It has been said that the role of the normal distribution in statistics is not unlike that of the straight line in geometry. It has very many uses, particularly in sampling theory of which quality control is just one special form. Integration enables us to discover some of the properties of the normal curve and so to apply these to statistical propositions.

And so the debt is repaid. At one time so little was known about curves that mathematicians developed the idea of treating them as approximate equivalents of rectilinear figures. The properties of the latter helped to measure the dimensions of the former. This in time led to the development of the calculus. Now we know so much about curves that in statistics we have been able to reverse the position. We now use the properties of curves to tell us something about the dimensions of rectilinear figures.

## Chapter 20

# PARTIAL DIFFERENTIATION AND INTEGRATION

In the previous chapters it has generally been assumed that the dependent variable $y$ is dependent only upon the variable $x$. But in real problems the variable $y$ may be dependent on more than one other variable. This of course introduces complications which are too advanced for detailed treatment in this chapter which is intended only as an introduction to the general ideas involved. The principles are also of value in the solution of differential equations even when there is only one independent variable involved.

If water flows through an inlet into a tank at the constant rate of $k$ gallons per hour then, after $x$ hours, the number of gallons of water in the tank ($y$) may be expressed in the function

$$y = kx$$

Here $y$ is a simple function of the variable $x$. Suppose however that the flow of water is not constant; let this be represented as $v$. In these circumstances

$$y = vx$$

where both $v$ and $x$ are variable.

When $y = kx$, we could calculate the derivative by first principles. Here $\frac{dy}{dx} = k$ because $k$ is constant and does not affect the variability in $y$. How can we calculate the derivative of $y = vx$, where the function has two independent variables? In this instance, the derivative is not equal to $v$ because $v$ is itself not a constant.

It is first assumed that the function is expressed in its simplest terms. That is, although $y$ depends upon $v$ and upon $x$, yet $v$ and $x$ are quite independent of each other. This assumption is necessary, for if $v$ itself depends upon $x$, then $y$ may also be expressed as a direct function of $x$. If, for example, $v = 4x$, then $y$ could be expressed entirely in terms of $x$

$$y = vx = 4x(x) = 4x^2$$

and the derivative could be calculated from first principles.

We therefore assume that we are dealing with two independent variables such that neither of these may be expressed as a function of the other. This is important since it follows that the variation in the value of one will not of itself produce any variation in the value of the other. Variation in $x$ therefore occurs irrespective of the value or even of the nature of $v$. Consequently the same variation in $x$ would occur even if $v$ were constant.

Since we may consider the variation in $x$ as quite distinct from any possible effect from $v$, then we may, in dealing with the variation in $x$, treat $v$ as if it really were constant. This will make no difference to the result. Similarly, in dealing with the variation in $v$, we may treat $x$ as constant.

It is necessary to be very clear about what we are doing here. Both $v$ and $x$ are variables and we treat *each* of these as if it were constant only while we are considering the variability in the *other*. We are *not* treating both variables as if they were really constant since this would obviously be incorrect.

When we treat $x$ as constant, the derivative $\frac{dy}{dv}$ will indicate the rate at which $y$ is varying because of its dependence on $v$. Similarly, when we treat $v$ as constant, the derivative $\frac{dy}{dx}$ will indicate the rate at which $y$ is varying because of its dependence upon $x$.

We thus have two partial derivatives neither of which measures the full amount of variation in $y$. Note that we specifically refer to the amount of variation. This can be measured. The rate of variation is another matter altogether.

Partial derivatives must of course be distinguished from ordinary or total derivatives. Thus, instead of writing $\frac{dy}{dx}$ as we do for the derivative of a function involving $x$ only, we use the notation $\frac{\partial y}{\partial x}$. The letter $d$ is replaced by the symbol $\partial$. This indicates that the derivative is a partial one only, as in the following example.

Let $$z = 3x^3+4x^2y+y^2$$

Here, $z$ is a function of $x$ and $y$. We first differentiate with respect to $x$, while $y$ remains constant. Then we differentiate with respect to $y$, while $x$ remains constant. That is:

(i) $\frac{\partial z}{\partial x} = 9x^2+8xy$ whence $dz_x = (9x^2+8xy)\,dx$

(ii) $\frac{\partial z}{\partial y} = 4x^2+2y$ ,, $dz_y = (4x^2+2y)\,dy$

In each result, the differential in $z$ relative to the differential in the independent variable involved is shown. The total differential in $z$ is the sum of these separate differentials and therefore

$$dz = (9x^2+8xy)\,dx+(4x^2+2y)\,dy$$

This is expressed in the general form as

$$dz = \frac{\partial z}{\partial x}.dx+\frac{\partial z}{\partial y}.dy$$

This formula provides a means of dealing with problems involving maximum or minimum values of functions of two variables. Such a problem is one in which it is necessary to specify the dimensions of an open top box of given volume so that the total surface area of its sides and base is a minimum.

Let $x$ = length of box; $y$ = width of box; $v$ = volume. Then the area of the base is $xy$ and the depth of the box is $\frac{v}{xy}$. With these dimensions, the total surface area is

$$A = xy+2x.\frac{v}{xy}+2y\frac{v}{xy} = xy+\frac{2v}{y}+\frac{2v}{x}$$

Differentiating with respect to $x$ gives

$$\frac{\partial A}{\partial x} = y-\frac{2v}{x^2} \quad \text{whence } dA_x = \left(y-\frac{2v}{x^2}\right)dx$$

Differentiating with respect to $y$ gives

$$\frac{\partial A}{\partial y} = x-\frac{2v}{y^2} \quad \text{whence } dA_y = \left(x-\frac{2v}{y^2}\right)dy$$

The total differential is the sum of the two partial differentials. Thus

$$dA = \left(y-\frac{2v}{x^2}\right)dx+\left(x-\frac{2v}{y^2}\right)dy$$

For a minimum value

$$\left(y-\frac{2v}{x^2}\right) = 0 \qquad \text{and also} \left(x-\frac{2v}{y^2}\right) = 0$$

Consequently $\quad 2v = yx^2 \qquad$ and also $2v = xy^2$

$$\therefore\ yx^2 = xy^2$$

$$\therefore\ x = y$$

The expression for the area may therefore be simplified by substituting $x$ for $y$. That is

$$A = xy+\frac{2v}{y}+\frac{2v}{x}$$

$$= x^2+\frac{2v}{x}+\frac{2v}{x} = x^2+\frac{4v}{x}$$

The area is now expressed as a function of only one variable (since $v$ is constant). We may therefore use the same procedure as in Chapter 7. That is, the derivative is equated to zero and the minimum value is thereby established.

$$\frac{dA}{dx} = 2x-\frac{4v}{x^2} = 0$$

$$\therefore\ 2x^3-4v = 0$$

$$\therefore\ x = \sqrt[3]{2v}$$

The depth of the box is $\frac{v}{xy}$ or $\frac{v}{x^2}$ (since $x = y$) and this may be simplified to $\frac{v}{\sqrt[3]{4v^2}}$ by substituting for $x$. All three dimensions are thus established for the given volume.

To obtain the total differential we merely add together the two partial differentials. This is made possible because each of the differentials measures the same kind of dimension and is expressed in the same units of measurement, just as we may add 1 hour to 1 hour and obtain a total of 2 hours.

We cannot however add 1 hour to 1 metre since the alleged total of 2 would be meaningless. For much the same reason we cannot add the two partial derivatives to obtain a total derivative if both of the variables in terms of which the function is expressed are also independent of each other. If the function $z$ is expressed in terms of both $x$ and $y$, where $x$ represents time and $y$ represents distance, then clearly the total rate of change in $z$ cannot be found by adding the rate per unit of time to the rate per unit of distance.

It is often the case that two variables while not being dependent on each other are nevertheless each dependent upon a third variable. This third variable is in many instances the variable of time, and in such circumstances it now becomes possible to calculate the total derivative which measures the rate of change of a function relative to time elapsed.

This results from the fact that if two variables $x$ and $y$ are functions of time (usually represented by $t$) then a function $z$ expressed in terms of $x$ and $y$ may also be expressed in terms of $t$. $z$ is then represented as a function of one variable $t$ and this function may be differentiated by basic principles, thus obtaining the derivative $\frac{dz}{dt}$.

Once we have calculated the partial derivatives, the rest is remarkably simple. This enables us to calculate the total differential $dz$. All that is then necessary is to divide this by $dt$ to obtain the total derivative, thus

$$dz = \frac{\partial z}{\partial x}.dx + \frac{\partial z}{\partial y}.dy$$

$$\therefore \frac{dz}{dt} = \frac{\partial z}{\partial x}\cdot\frac{dx}{dt} + \frac{\partial z}{\partial y}\cdot\frac{dy}{dt}$$

As a simple example we take the function $z = x^2 + y^3$ where $x = 3t$ and $y = 5t$. The procedure is as follows:

(i) $x = 3t$ whence $\frac{dx}{dt} = 3$

(ii) $y = 5t$ ,, $\frac{dy}{dt} = 5$

(iii) $\dfrac{\partial z}{\partial x} = 2x$

(iv) $\dfrac{\partial z}{\partial y} = 3y^2$

$$\therefore \frac{dz}{dt} = (2x)(3)+(3y^2)(5)$$
$$= 6x+15y^2$$

This may be checked by substituting for $x$ and $y$ in the original function.

$$z = x^2+y^3$$
$$= (3t)^2+(5t)^3$$
$$= 9t^2+125t^3$$
$$\therefore \frac{dz}{dt} = 18t+375t^2$$
$$= 6(3t)+15(5t)^2$$
$$= 6x+15y^2$$

The above is an intentionally simple example. Where the function is a complicated one then the method of substituting for $x$ and $y$ will also involve some complicated working; partial differentiation will usually be found to offer a simple method. Further reference is made to partial differentiation in connection with the solution of differential equations in Chapter 22, and a useful application for a total differential is noted in Chapter 21.

## PARTIAL INTEGRATION

Just as integration is the reverse of differentiation, so is partial integration the reverse of partial differentiation. Thus, if we have an integrand containing two variables, we first integrate as if one of the variables were constant. The result of this integration is then integrated, but this time the other variable is treated as if it were constant. This might be expressed in the following form:

$$z = \iint (x^2+y^2)\, dy\, dx$$

The double integral sign means that we have to integrate twice, and the two differential signs $dy$ and $dx$ indicate that we are required to integrate once with respect to $y$ and once with respect to $x$.

If this is first integrated with respect to $y$ (treating $x$ as constant) we obtain

$$x^2y+\frac{y^3}{3}+c_1$$

$c_1$ is the constant of integration. But since we are treating $x$ as constant then $c_1$ may in fact involve $x$. For example, if $c_1 = 3$ or $3x$ it would still be treated as constant. Hence it is usual to replace $c_1$ by some symbol as $\phi(x)$ which indicates that $\phi$ depends upon $x$.

This expression is now integrated with respect to $x$ (treating $y$ as constant). This gives the final integral as

$$z = \frac{x^3y}{3}+\frac{xy^3}{3}+\phi_1(x)+\psi(y)$$

The constant of integration arising in this second integration is shown as $\psi(y)$ to indicate its dependence upon $y$. It will also be noted that the constant depending on $x$ is now shown as $\phi_1(x)$ to indicate that this too has been integrated. That is

$$\phi_1(x) = \int \phi(x)\,dx$$

It will be seen that in the indefinite form the appearance of constants may bring some complication, but as in other forms of integration the constants will disappear when appropriate limits are applied to the variables.

*Areas by double integration*

In Chapter 9, we were able to calculate the area under a curve by equating it to the limiting value of the sum of an indefinite number of trapeziums:

$$\text{Area} = \int y\,dx$$

and it was noted that this was numerically equivalent to the limiting value of the sum of rectangles $y$ by $\Delta x$, as $\Delta x$ is made progressively smaller.

In Fig. 64 is an illustration that the area of each of these trapezium strips may itself be regarded as being equal to the limiting value of the sum of the areas of rectangles $\Delta x$ by $\Delta y$. That is, the area of each strip is

$$A_s = \int dy\,dx$$

The area under the curve is the limiting value of the sum of the areas of the strips and is therefore

$$A = \iint dy\, dx$$

In this example the limits for $x$ are $a$ and $b$. The limits for $y$ are 0 and $f(x)$, the function of $x$ represented by the curve. Consequently the full expression for the total area under the curve is

$$\int_a^b \int_0^{f(x)} dy\, dx$$

The inner integral sign shows the limits of $y$ for the integration of the rectangles within the strips; the outer integral sign shows the limits of $x$ for the integration of the strips themselves.

This process may be illustrated as follows, in the calculation of the area under the curve of $y = x^2$. This area, calculated by basic integration principles, is $\frac{1}{3}x^3$. The result is now obtained by partial integration. The derivative of this function is

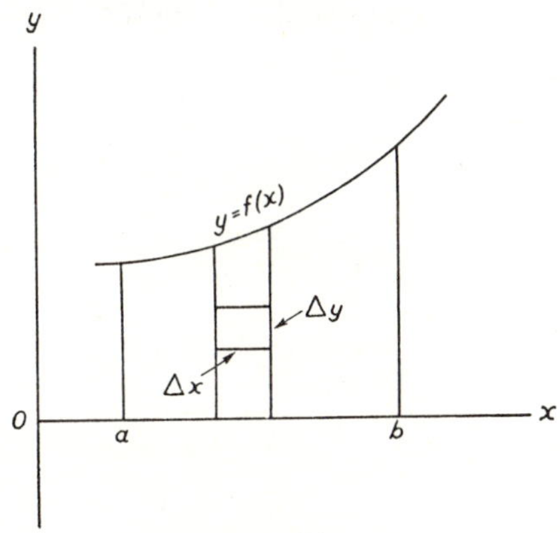

FIG. 64

$$\frac{dy}{dx} = 2x$$

and $$dy = 2x\, dx$$

Consequently $$dy\, dx = 2x\, dx\, dx$$

and $$\int dy\, dx = \int 2x\, dx\, dx$$

and $$\iint dy\, dx = \iint 2x\, dx\, dx = \text{Area}$$

$$= \int x^2\, dx$$

$$= \tfrac{1}{3}x^3 + \phi_1(x) + \psi(y)$$

*Volumes by triple integration*

The basic idea demonstrated above may be extended to the calculation of the volumes of a solid. For this purpose it is necessary to integrate three times.

The solid is regarded as consisting of an indefinite number of slices of width $dx$. Each slice is regarded as consisting of columns, and each column is regarded as consisting of rectangular parallelepipeds of dimensions $dx$, $dy$, and $dz$.

To calculate the volume we first integrate the volume of the parallelepipeds into columns; the columns are then summed to give the volume of the slices and the slices are then summed to give the total volume of the solid. This is expressed as

$$v = \iiint dz\, dy\, dx$$

*Chapter 21*

# CURVATURE

Every motorist and other road-user will know that some roads 'bend' or curve more sharply than others, just as some hill roads climb more steeply than others. The steepness of a hill is conveniently measured by its gradient. The bending of a road or of a curve is measured by its curvature which relates the amount of change in direction of the curve to its length. A curve which moves gently away from a straight line has a lower curvature ratio than has a curve which moves away sharply.

The amount of curvature in a road or in a railway track is of great practical importance for the safety of vehicles which travel along them. A long slow turn in a road is easier for a car to negotiate than is a short sharp turn. If a car enters a turn in the road at a speed which is too high for the curvature of the road it is likely to overturn or to run off the road. A train entering a curve at speed will jump the rails if the curve is too sharp.

This result arises from the fact that when a vehicle enters the curve it becomes subject to forces the effect of which is to try to make the vehicle continue in a straight line of direction. Other forces have to be applied to the mass of the vehicle to make it enter the curve. Failure to allow for the effect of these forces would be disastrous. This in part explains why some motor-race courses, roads, and railway tracks are banked so that their surfaces slope down towards the centre of a figure of which the curved track may be regarded as part of the perimeter. This banking sets up opposing forces and allows the curve to be made more sharply than would otherwise be possible.

The 'safety' of a curve thus depends partly upon its curvature, as well as upon the behaviour of the vehicle. In order to decide what is a safe form of curve we must therefore be able to measure its curvature.

If a curve may be regarded as the arc of a circle, then the force exerted on a moving vehicle is given by

$$F = \frac{mv^2}{r}$$

where $m$ = mass of vehicle, $v$ = velocity, and $r$ = radius of the circle of which the curve is an arc. From this it will be seen that $F$ is decreased as $r$ is increased. For a given value of $mv^2$, the force may be decreased by increasing the radius of the circle.

The larger the circle then, provided $mv^2$ remains constant, the lower the force. Similarly, the larger the circle, so the less is its curvature along an arc of given length. This can be seen in Fig. 65 which shows two circles of radius $r$ and $R$ respectively. The lengths

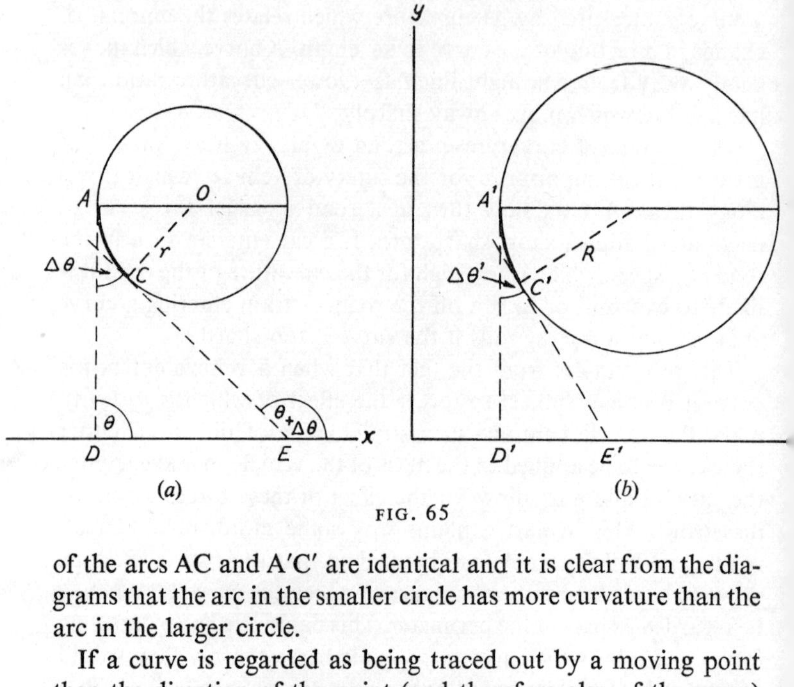


FIG. 65

of the arcs AC and A'C' are identical and it is clear from the diagrams that the arc in the smaller circle has more curvature than the arc in the larger circle.

If a curve is regarded as being traced out by a moving point then the direction of the point (and therefore also of the curve) changes as the point moves. As the point starts from A in the smaller circle it moves in the direction AD. When it reaches C it is moving in the direction CE.

The amount of change of direction for a small piece of arc is measured by the angle $\Delta\theta$ (in radians) (and it will be noted that this measures also the increment in the angle $\theta$ enclosed by the line of direction and the $x$ axis). The angle therefore measures the *total curvature* of the respective arc and it is evident that $\Delta\theta'$ is less than $\Delta\theta$. The *average curvature* is calculated by dividing the total curvature by the length of the arc. The increment in the length of the arc is usually represented by $\Delta s$. Accordingly the average curvature of this increment is

$$\frac{\Delta\theta}{\Delta s}$$

As $\Delta s$ is made smaller, so this ratio usually approaches a limiting value at point C. The curvature ($k$) of the arc *at a point* is therefore defined as

$$k = \lim_{\Delta s \to 0} \left[\frac{\Delta\theta}{\Delta s}\right] = \frac{d\theta}{ds}$$

To calculate $k$ for any point on a circle we proceed as follows. The angle $\Delta\theta$ is enclosed between tangents drawn at two points on the circumference of the circle. It is a general property of circles that such an angle is always equal to the angle enclosed at the centre of the circle by the radii drawn at the same points. Thus, angle $AOC = \Delta\theta$.

Furthermore the length of an arc of a circle may be calculated by multiplying the length of the radius by the numerical value (in radians) of the central angle subtended by the arc.

Thus $$\Delta s = \Delta\theta\, r$$

or $$\frac{\Delta\theta}{\Delta s} = \frac{1}{r}$$

This is a constant value and the limiting form of $\frac{\Delta\theta}{\Delta s}$, that is $\frac{d\theta}{ds} = \frac{1}{r}$.

The circle is a curve of constant curvature and the ratio $\frac{1}{r}$ symbolizes the fact that the greater the radius of the circle, so the less is its curvature. It also shows that the curvature at any point on the circle is the same as the average curvature of an arc. If the radius is 1 unit, then $k = \frac{1}{r} = 1$ (in radians) for a unit length of arc (i.e. an arc equal in length to the radius).

The procedure is not so easy with other curves since these do not have constant curvatures. Fig. 66 shows a curve drawn relative to a system of co-ordinate axes.

AC is a small part of the arc. Let AC $= \Delta s$. Also in the diagram is a circle which is drawn so that a very small part of its circumference almost coincides with arc AC. The smaller the arc so the more easily should we be able to find such a circle, the idea being that we may regard the curvature of the circle at a point as being the same as the curvature of the arc at the same point. This circle is referred to as the circle of curvature or the osculating circle at point A. The radius of the circle is called the *radius of curvature*

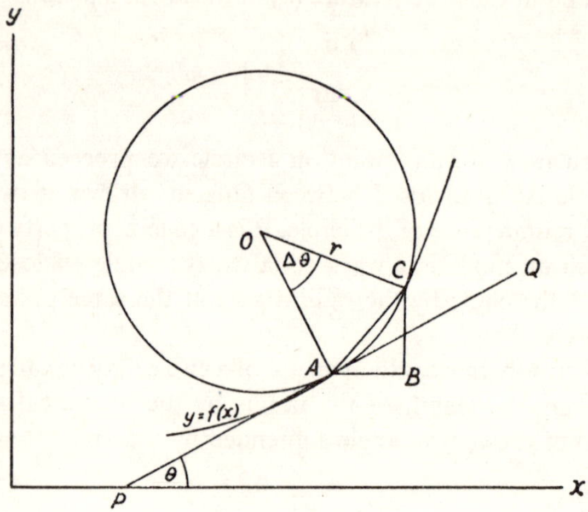


FIG. 66

of the curve at that point. As $\Delta s$ is made progressively smaller then the curvature of this piece of arc will approach equality with the curvature of the circle. Thus, while the equation $\frac{\Delta\theta}{\Delta s} = \frac{1}{r}$ is approximately true, the equation $\frac{d\theta}{ds} = \frac{1}{r}$ is *exact*.

If we could construct the circle then the value of $\frac{1}{r}$ could be obtained from it. But of course we need to know the value of $\frac{1}{r}$ before we can construct the circle! Some other way must be found.

As $\Delta s$ approaches zero, the secant AC approaches the tangent at A, which meets the $x$ axis at the angle $\theta$. The slope of this tangent is $\tan\theta = \frac{dy}{dx}$. This in turn may be differentiated to give the second derivative with respect to $x$ $\left(\text{that is, } \frac{d^2y}{dx^2}\right)$.

The derivative of $\tan\theta$ *with respect to* $\theta$ is $\sec^2\theta$ (see Chapter 13). We need the derivative *with respect to* $x$ and we therefore multiply $\sec^2\theta$ by $\frac{d\theta}{dx}$. That is

$$\frac{d^2y}{dx^2} = \frac{d(\tan\theta)}{d\theta} \times \frac{d\theta}{dx} = \sec^2\theta \times \frac{d\theta}{dx}$$

or, since $$\sec\theta = \frac{1}{\cos\theta}$$

$$\frac{d^2y}{dx^2} = \frac{1}{\cos^2\theta} . \frac{d\theta}{dx}$$

whence $$\frac{d\theta}{dx} = \cos^2\theta . \frac{d^2y}{dx^2}$$

Now $\frac{1}{r} = \frac{d\theta}{ds}$ and to obtain this we multiply $\frac{d\theta}{dx}$ by $\frac{dx}{ds}$. But $\frac{dx}{ds} = \cos\theta$. This is obtained as follows. In the triangle ACB (Fig. 66) the ratio $\frac{\text{AB}}{\text{AC}}$ is equal to cos CAB. As C is moved towards A, the angle CAB approaches the angle QAB which is equal to $\theta$. Therefore cos CAB approaches cos QAB or $\cos\theta$; at the same time the ratio of arc AC to chord A Capproaches 1, whence $\frac{\text{AB}}{\text{arc AC}}$ also approaches $\cos\theta$. That is

$$\frac{\Delta x}{\Delta s} \to \cos\theta \qquad \text{and} \qquad \frac{dx}{ds} = \cos\theta$$

Then $$\frac{1}{r} = \frac{d\theta}{ds} = \frac{d\theta}{dx} \times \frac{dx}{ds} = \cos^2\theta . \frac{d^2y}{dx^2} . \cos\theta$$

$$= \cos^3\theta . \frac{d^2y}{dx^2}$$

This gives an expression for $\frac{1}{r}$ but it still contains the variable $\theta$.

To remove this we note the equivalents (see Appendix II):

$$\cos\theta = \frac{1}{\sec\theta}$$

and $$\sec\theta = \pm\sqrt{1+\tan^2\theta}$$

For reasons given below we take the positive root here and substituting in the above expression we obtain

$$\frac{1}{r} = \frac{1}{(\sqrt{1+\tan^2\theta})^3}\cdot\frac{d^2y}{dx^2}$$

But we know that $\tan\theta = \dfrac{dy}{dx}$ and therefore

$$\frac{1}{r} = \frac{1}{\left[1+\left(\dfrac{dy}{dx}\right)^2\right]^{\frac{3}{2}}}\cdot\frac{d^2y}{dx^2}$$

If we use the abbreviated symbolism for derivatives then

$$\frac{1}{r} = \frac{f''(x)}{[1+(f'(x))^2]^{\frac{3}{2}}}$$

and $r$, the radius of curvature, is of course the reciprocal of this

$$r = \frac{[1+(f'(x))^2]^{\frac{3}{2}}}{f''(x)}$$

The numerator involves a square root which may be either positive or negative and for this purpose the root must always have the same denominator so that $r$ is always positive.

The centre of the circle of radius $r$ is called *the centre of curvature*, and we must be able to locate this point relative to the $x$ and $y$ axes. The value of $r$ tells us the radius of the circle but gives no direct indication of the position of the centre.

Let the centre of the circle be represented by the point $(x_1, y_1)$. When a circle is placed so that its centre coincides with the origin of the co-ordinate axes, its equation is

$$x^2+y^2 = r^2$$

The present circle is differently placed however, and in its present position its equation is modified to

$$(x-x_1)^2+(y-y_1)^2 = r^2$$

This is illustrated in Fig. 67 where OB $= (x-x_1)$; OC $= r$; and BC $= (y-y_1)$.

The equation may be expanded to

$$(x^2-2x_1x+x_1^2)+(y^2-2y_1y+y_1^2) = r^2$$

In this $r$ is constant and so are $x_1$ and $y_1$ constant for the specific circle in which we are interested. This equation therefore contains two variables. Before we can hope to identify $x_1$ and $y_1$ we must

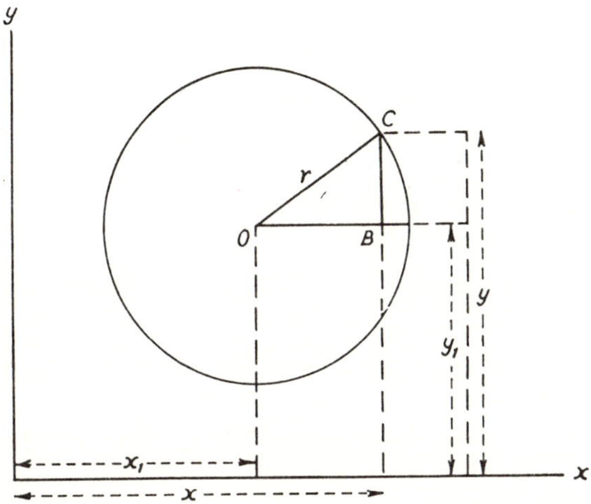


FIG. 67

dispose of the constant $r$. Since the equation is an identity for all values representing points of the circle, so that one side of the equation is always equal to the other, then the results of differentiating each side must also be equal to each other. Partial differentiation is required here since both $x$ and $y$ are variable. If we differentiate the left-hand side of the equation first with respect to $x$ and then with respect to $y$ we obtain the total differential

$$(2x-2x_1)\,dx+(2y-2y_1)\,dy$$

Since $r$ is constant the total differential of $r^2$ is 0 (that is 0 with respect to $x$, plus 0 with respect to $y$). Then

$$(2x-2x_1)\,dx+(2y-2y_1)\,dy = 0$$

Now divide by $2dx$:

$$x-x_1+(y-y_1)\frac{dy}{dx}=0 \qquad \text{(i)}$$

This is now differentiated with respect to $x$. This is not a straightforward matter. There are three terms to consider

(*a*) The derivative of $x$ $= 1$
(*b*) „ „ „ $x_1$ (a constant) $= 0$
(*c*) The third term is a product

$$(y-y_1)\times\frac{dy}{dx}$$

and we need the product formula for differentiation.

This formula shows that if $f(x) = uv$, then

$$f'(x) = u.\frac{dv}{dx}+v.\frac{du}{dx}$$

In our example we have $f(x) = (y-y_1)\frac{dy}{dx}$, where $u = (y-y_1)$ whence $\frac{du}{dy} = 1 \left(\text{so } du = dy \text{ and, therefore, } \frac{du}{dx} = \frac{dy}{dx}\right)$. Also $v = \frac{dy}{dx}$. Substituting these equivalents in the formula gives the derivative as

$$(y-y_1)\left(\frac{d^2y}{dx^2}\right)+\left(\frac{dy}{dx}\right)^2$$

Differentiating (i) therefore gives

$$1+\left(\frac{dy}{dx}\right)^2+(y-y_1)\frac{d^2y}{dx^2}=0 \qquad \text{(ii)}$$

This differentiation is possible since equation (i) is an identity, holding good for every value of the variables contained in it. Before proceeding further however we should refer to a point which sometimes causes difficulty. We have seen that we can differentiate the left side of the equation. We then equate the derivative of the left side to the derivative of the right side. But the right side of the equation consists only of 0. How can we differentiate 0? Surely the derivative of nothing cannot exist? If this is true, then the 'differentiation' is impossible.

The fact that the differentiation is valid however may be shown quite easily. Equation (i) is merely a particular form of expressing the relationship between its constituent variables. In effect it says that the left side of the equation may be split into two equal parts so that the subtraction of one from the other gives a result equal to zero. The equation could just as easily be transformed to

$$x+(y-y_1)\frac{dy}{dx} = x_1$$

merely by transposing $x_1$, and changing its sign. We no longer have to contend with 0. Each side may be differentiated giving exactly the same result as in equation (ii), since the derivative of $x_1$ is 0.

Simplifying and transposing equation (ii) gives

$$y_1 = y + \frac{1+\left(\dfrac{dy}{dx}\right)^2}{\dfrac{d^2y}{dx^2}}$$

Substituting this value in equation (i) and then simplifying also gives

$$x_1 = x - \frac{\dfrac{dy}{dx}\left[1+\left(\dfrac{dy}{dx}\right)^2\right]}{\dfrac{d^2y}{dx^2}}$$

and the position of the centre of the circle is thus definitely fixed.

All these calculations may now be applied to a specific curve. If

$$y = 3x^2-x+4$$

what is the length of the radius of curvature ($r$); the curvature of the curve $\left(\dfrac{1}{r}\right)$; and the co-ordinates of the centre of curvature $(x_1, y_1)$ when $x = 0$?

First calculate the derivatives

$$\frac{dy}{dx} = 6x-1 \qquad \frac{d^2y}{dx^2} = 6$$

*Radius*

Then $$r = \frac{[1+(6x-1)^2]^{\frac{3}{2}}}{6}$$ (see page 274)

When $x = 0$, this becomes

$$r = \frac{\sqrt{2^3}}{6} = \frac{\sqrt{2}}{3}$$

*Curvature*

Then $$\frac{1}{r} = \frac{3}{\sqrt{2}}$$

*Co-ordinates*

Substituting the equivalents of the derivatives in the equation for $x_1$ gives

$$x_1 = x - \frac{[(6x-1)][1+(6x-1)^2]}{6}$$

when $x = 0$, this becomes $\frac{1}{3}$.

Similarly substituting the equivalents of the derivatives in the equation for $y_1$ (and noting from the original function that $y = 4$ when $x = 0$) gives

$$y_1 = y + \frac{1+(6x-1)^2}{6}$$

$$= 4+\tfrac{1}{3} = 4\tfrac{1}{3}$$

These results may be easily checked since they must satisfy the equation of the circle

$$(x-x_1)^2+(y-y_1)^2 = r^2$$

Substituting the results obtained gives

$$(0-\tfrac{1}{3})^2+(4-4\tfrac{1}{3})^2 = \left(\frac{\sqrt{2}}{3}\right)^2$$

or $$\tfrac{1}{9}+\tfrac{1}{9} = \tfrac{2}{9}$$

The results therefore do satisfy the equation.

# Chapter 22

# ELEMENTARY DIFFERENTIAL EQUATIONS

In the demonstrations of the processes of integration we have encountered a number of equations involving derivatives or differentials. These are called differential equations. Their solution consists in establishing the relationship between the independent and dependent variables involved, and this is achieved by integration.

From what we know about the difficulties of integration it is at once obvious that not all differential equations are completely soluble since there are some expressions which cannot be integrated. Consequently it is not possible to formulate a set of rules which will apply to each and every equation.

This does not mean that technically insoluble equations are of no value. The equations which arise in the study of physical phenomena express the relationships between the variables involved. While it may not be possible to solve them completely it may be possible to obtain limited results, as for example when one effect expressed therein may be regarded as negligible. The equations may also assist in developing physical theories even though they cannot be solved numerically.

Here we are concerned only with those equations which can be integrated. Even these may at first glance appear to be of such a form as to make integration impossible and it is the object of this chapter to consider methods of dealing with equations in general so as to obtain solutions. The emphasis is not so much on integration itself as upon ways of transforming equations so as to make integration possible.

First, a few notes about the general character of equations. An *ordinary* differential equation involves only one independent variable, a *partial* differential equation involves more than one independent variable.

Differential equations are classified in *orders* according to the highest derivative which occurs in them. An equation which contains a first derivative is of the first order whereas one which contains a second derivative is of the second order. Equations are further classified in *degrees*, according to the highest power of the highest derivative included. If, for example, an equation contains both a first derivative and the cube of a second derivative, then the equation is of the third degree. The highest derivative is the second derivative and the exponent of this is three.

When a function is integrated an arbitrary constant appears. Where successive integration is performed a constant appears on each integration. Consequently the number of constants which will appear in the general solution of an equation is equal to the number representing the order of the equation. Since these constants are arbitrary and we may assign any values to them without affecting the correctness of the solution it follows that the equation has an infinity of solutions.

A *general solution* is one which includes the symbols for the arbitrary constants but does not assign numerical values to them. In a particular problem however these constants will have a special meaning, since they may already be known or may perhaps be derived in certain instances from the equation itself in association with observed results for given related values of the independent and dependent variables. When the actual values of these constants are substituted in a general solution the latter is then called a *particular solution*.

For the purposes of solving equations it is possible and convenient to identify different types of equation and to develop a method suitable for each respective type. We now proceed to consider certain types of ordinary equations.

A differential equation may be regarded as having been derived by differentiation from an original equation which expresses the functional relationship between the variables. This original equation is referred to as the primitive of the differential equation. The latter is solved when we have identified its primitive.

The simplest type of equation is of the form

$$\frac{dy}{dx} = f(x)$$

which merely requires ordinary integration. Thus if

$$\frac{dy}{dx} = 3x^2$$

then $$dy = 3x^2\,dx$$

and $$y = \int dy = \int 3x^2\,dx$$
$$= x^3 + c$$

In the above equation the derivative was expressed as a function of $x$. In some equations it may be expressed as a function of $y$ as, for example, in

$$\frac{dy}{dx} = \tan y$$

If we transform this to

$$dy = \tan y\,dx$$

we have on the right side an expression which cannot be integrated. Instead therefore we transform the equation to

$$dx = \frac{dy}{\tan y}$$

and then by ordinary integration

$$x = \int dx = \int \frac{dy}{\tan y}$$
$$= \int \frac{\cos y}{\sin y}.dy$$
$$= \log_e \sin y + c_1$$

The result follows from the rule for the integration of a fraction the numerator of which is the derivative of the denominator. The integral of the fraction is then equal to the logarithm of the denominator (see page 148).

## EQUATIONS WITH SEPARABLE VARIABLES

In the above examples we had equations in which only one of the variables appeared. In many equations however both the independent and the dependent variables appear. If the terms of such an equation may be arranged in two groups so that each group contains only one variable, then the variables are said to be separable. The terms as they appear in the original equation may not be immediately separable and it is then necessary to transform the equation to make integration possible.

As an example we take the equation

$$x\,dy+y\,dx = 0$$

At first the variables do not appear to be separable. Yet they may in fact be separated by multiplying the equation throughout by $\frac{1}{xy}$ (or, which is the same thing, dividing by $xy$). Then

$$\frac{x\,dy}{xy}+\frac{y\,dx}{xy} = 0$$

whence

$$\frac{dy}{y}+\frac{dx}{x} = 0$$

and integration is now possible

$$\int\frac{dy}{y}+\int\frac{dx}{x} = c_1$$

The integral of $\int\frac{dx}{x}$ is the same as $\int\frac{1}{x}.dx$ and this integral is equal to $\log_e x$ (see page 146). Similarly $\int\frac{dy}{y} = \log_e y$.

Then

$$\log_e y+\log_e x = c_1$$

But, by the rules of logarithms

$$\log_e y+\log_e x = \log_e xy$$

$$\therefore \log_e xy = c_1$$

If $\log_e xy$ is constant, then $xy$ must also be constant

$$\therefore xy = c$$

The factor $\frac{1}{xy}$, which enabled the original equation to be transformed so that the variables could be separated and so that the separate terms could be integrated, is called the *integrating factor*. The factor required for this purpose will of course depend upon the particular equation to be solved.

## GENERAL FORMS OF EQUATIONS

When the separate terms of the above equation were integrated we appear to have integrated the right-hand side of the equation

(0) to $c_1$. Is this true or even possible? How can we integrate – build-up – anything at all from nothing?

There are two ways of thinking about this. Firstly, we might note that it is true that the derivative of a constant is zero. Integration is the reverse of differentiation and, accordingly, if the derivative of a quantity is 0 then that quantity must be a constant. Thus perhaps we could argue that as

$$\frac{dy}{dx} = 0$$

then
$$dy = 0.dx$$

$$y = \int 0.dx = c_1$$

The difficulty about this however is that the derivative is zero because $dy$ is zero in the first place and we still seem to be playing about with a lot of nothing.

It is better in fact not to try to think of $\int 0.dx$ being integrated in the usual sense. Fortunately there is no need to do so. The zero symbol in the original equation has no meaning other than signifying the 'emptiness' of the right-hand side of the equation when all the terms have been collected on the left-hand side.*

The equation could also have been expressed as

$$x\,dy = -y\,dx$$

and it could have been solved by using the same method without transferring the right-hand term to the left side of the equation. Thus, using the same integrating factor,

$$\frac{dy}{y} = -\frac{dx}{x}$$

$$\therefore \int \frac{dy}{y} = -\int \frac{dx}{x}$$

$$\therefore \log_e y + c_y = -\log_e x + c_x$$

$$\therefore \log_e y + \log_e x = c_x - c_y \; (= c_1)$$

In other words it makes no difference at what stage we transfer terms from one side of the equation to the other; the result is the

* See also page 276.

same. In this latter method of working however the nett constant of integration arises naturally from the integration of the separate terms neither of which is equal to zero.

It is nevertheless usual for this particular type of equation to be expressed in the form in which all the terms appear on the left side of the equation. The effect of both equations:

$$x\,dy = -y\,dx$$

and

$$x\,dy+y\,dx = 0$$

is the same, but the second form assists in the classification of equations. The terms are arranged so that we may more clearly see what they are.

This has a special significance in the present instance. We obtained the solution

$$xy = c$$

If we apply the process of partial differentiation to this equation (see Chapter 20) then, treating $x$ as variable, $y\,dx = 0$. Similarly, treating $y$ as variable, then $x\,dy = 0$. Consequently the total differential

$$x\,dy+y\,dx = 0$$

It is seen therefore that the differential equation to be solved may be regarded as having been derived by partial differentiation from the primitive

$$xy = c$$

and that this provides the desired solution. Recognition that part of this equation is in the form of a total differential provides the clue as to how the solution may be achieved.

The general form of equations of this type is

$$M\,dx+N\,dy = 0$$

where $M$ and $N$ are functions of $x$ or $y$ or both. An equation of this form is called an exact equation if it is formed from its complete primitive by differentiation.

A general rule has been formulated for solving exact differential equations but before this can be used it is first necessary to ensure that an equation to be solved is in fact exact. Fortunately there is a relatively simple test which may be used to check for exactness.

This test depends upon the properties of partial derivatives, and may be developed as follows:

The primitive must be of the form

$$f(x, y) = u$$

That is, the left-hand side must be a function of $x$ and $y$ since if the equation

$$M\,dx + N\,dy = 0$$

has been obtained from the primitive then the latter has been differentiated first with respect to $x$ and then with respect to $y$. The differential $M\,dx$ is obtained from the differentiation with respect to $x$. That is, $M$ is the partial derivative of the primitive differentiated with respect to $x$.

$$M = \frac{\partial u}{\partial x}.$$

Similarly $N$ is the partial derivative of the primitive differentiated with respect to $y$.

$$N = \frac{\partial u}{\partial y}$$

The equation is therefore of the form

$$\frac{\partial u}{\partial x}.dx + \frac{\partial u}{\partial y}.dy = 0$$

including the first partial derivatives in place of $M$ and $N$.

These partial derivatives are themselves functions of the variables involved in the primitive and may therefore also be partially differentiated to give higher partial derivatives of the primitive. This will result in four second partial derivatives since each of the first partial derivatives may be differentiated both with respect to $x$ and with respect to $y$. These are shown in the following table.

| *First partial derivative* | *Differentiate with respect to* | *Second partial derivative* |
| --- | --- | --- |
| $\frac{\partial u}{\partial x}$ | $x$ | $\frac{\partial}{\partial x}\left(\frac{\partial u}{\partial x}\right) = \frac{\partial^2 u}{\partial x^2}$ |
| | $y$ | $\frac{\partial}{\partial y}\left(\frac{\partial u}{\partial x}\right) = \frac{\partial^2 u}{\partial y\,\partial x}$ |
| $\frac{\partial u}{\partial y}$ | $x$ | $\frac{\partial}{\partial x}\left(\frac{\partial u}{\partial y}\right) = \frac{\partial^2 u}{\partial x\,\partial y}$ |
| | $y$ | $\frac{\partial}{\partial y}\left(\frac{\partial u}{\partial y}\right) = \frac{\partial^2 u}{\partial y^2}$ |

Of these second partial derivatives it can be shown that those on the second and third lines of the table are in fact equal to each other. The difference in the order of the symbols $\partial x$ and $\partial y$ merely indicates the order in which the successive partial differentiations have been carried out but this has no effect on the value of the second partial derivative obtained. Thus if we differentiate first with respect to $x$ and then differentiate the partial derivative with respect to $y$, we shall obtain the same result as if we had first differentiated with respect to $y$ and then differentiated the partial derivative with respect to $x$.

Thus
$$\frac{\partial^2 u}{\partial y\, \partial x} = \frac{\partial^2 u}{\partial x\, \partial y}$$

But $M = \frac{\partial u}{\partial x}$ and $N = \frac{\partial u}{\partial y}$. Substituting these in the above equation gives

$$\frac{\partial}{\partial y}.M = \frac{\partial}{\partial x}.N$$

and this gives the basis for the test of whether the differential equation to be solved is exact. All we need do is to differentiate $M$ with respect to $y$, and then to differentiate $N$ with respect to $x$. The two results thus obtained will be equal to each other if the equation is exact but not otherwise.

As an example we take the differential equation

$$(3x^2+8xy)\, dx+(4x^2+3y^2)\, dy = 0$$

Is this exact? In this instance

$$M = 3x^2+8xy \qquad \text{whence } \frac{\partial}{\partial y}.M = 8x$$

$$N = 4x^2+3y^2 \qquad \text{whence } \frac{\partial}{\partial x}.N = 8x$$

The results are equal and the equation is therefore exact.

To obtain the primitive of an exact equation (that is, to solve the equation) we remember that $M$ and $N$ are regarded as having been obtained by partial differentiation and accordingly the solution is obtained by reversing the process; that is, by integration. We therefore calculate the integrals $\int M\, dx$ and $\int N\, dy$ and then, with one very important reservation, the sum of these two integrals represents the left-hand side of the solution. This reservation is that any terms common to both integrals will appear only *once* in the final solution.

In the above example

$$\int M\,dx = \int (3x^2+8xy)\,dx = x^3+4x^2y$$

$$\int N\,dy = \int (4x^2+3y^2)\,dy = 4x^2y+y^3$$

The term $4x^2y$ appears in each integral but it appears only once in the left-hand side of the solution which is

$$x^3+4x^2y+y^3$$

In considering the primitive function we assumed that it was of the form

$$f(x, y) = u$$

Consequently the restoration of the function $f(x, y)$ by integration must be accompanied by the restoration of its equivalent $u$.

The full solution of the equation is therefore

$$x^3+4x^2y+y^3 = c$$

where $c$ represents the constant $u$.

It has been said that, in building up the solution from $\int M\,dx$ and $\int N\,dy$, terms common to both are included once only. The reason for this may require clarification. When we differentiate the primitive with respect to $x$, every term involving $x$ will have a corresponding term in the total differential while those involving $y$ only will disappear ($y$ being treated as constant). On the other hand when we differentiate with respect to $y$, all terms involving $y$ in the primitive will have corresponding terms in the total differential.

This means that if a term in the primitive involves $x$ *and* $y$, it will have two corresponding terms in the differential. Upon integration each of these two terms will reinstate the original term of the primitive so that this term is reinstated twice, whereas of course it actually appears in the primitive only once. This duplication of terms has to be cancelled. It is a kind of duplication which is quite common. For instance if there are two boys and each has a father then there are two fathers, unless the boys are brothers when there is only one father. If we regard the terms in a differential equation as the offspring of the terms in the primitive then the need to avoid duplications will be more clearly recognized.

If an equation of this general form is not exact it may be possible to transform it into a form which is exact. This is achieved with the help of an integrating factor but it is not always an easy matter to discover the appropriate factor for any particular equation. This procedure makes it possible to find some solutions even in instances where the variables are not separable.

The equation

$$4y\,dx+x\,dy = dx$$

cannot be integrated in its present form. The left-hand side is not in the form of a total differential. Multiplying by the integrating factor $x^3$ however produces an equation which can be integrated

$$4x^3y\,dx+x^4\,dy = x^3\,dx$$

The left-hand side of this equation is exact. (Check: $4x^3y$ differentiated with respect to $y$ gives $4x^3$; while $x^4$ differentiated with respect to $x$ also gives $4x^3$.)

$$\int 4x^3y\,dx = x^4y$$

$$\int x^4\,dy = x^4y$$

$$\int x^3\,dx = \frac{x^4}{4}$$

The solution of the equation is therefore

$$x^4y = \frac{x^4}{4}+c$$

## HOMOGENOUS EQUATIONS

These equations are of the general form

$$M+N\frac{dy}{dx} = 0$$

where $M$ and $N$ are homogenous functions of the same degree throughout in $x$ and $y$. Equations of this form may be solved by use of the substitution $\frac{y}{x} = v$ (or $y = vx$).

In the equation

$$(x+y)+x.\frac{dy}{dx} = 0$$

we have $M = x+y$; $N = x$. $M$ and $N$ are homogenous functions. Let $y = vx$. Then, by the product formula,

$$dy = v\,dx+x\,dv$$

whence

$$\frac{dy}{dx} = \frac{v\,dx+x\,dv}{dx}$$

Substitute for $y$ and $\frac{dy}{dx}$ in the original equation giving the transform

$$(x+vx)+x.\frac{v\,dx+x\,dv}{dx} = 0$$

Multiply this throughout by $\frac{dx}{x}$ and simplify to

$$dx+2v\,dx+x\,dv = 0$$

$$\therefore\ (1+2v)\,dx+x\,dv = 0$$

Separate the variables by multiplying by the integrating factor $\frac{1}{(1+2v)\,x}$

$$\frac{dx}{x}+\frac{dv}{1+2v} = 0$$

Integrate the terms separately

$$\log_e x+\tfrac{1}{2}\log_e (1+2v) = c_1 \qquad \text{(A)}$$

This result follows from the rule for the integration of a fraction the numerator of which is the derivative of the denominator; the integral of the fraction is then equal to the logarithm of the denominator.

The fraction in the integrand $\frac{1}{1+2v}.dv$ is not of this form since the derivative of $1+2v$ is 2, whereas the numerator is 1. Nevertheless this may be transformed to $\left(\frac{1}{2}\right)\left(\frac{2}{1+2v}\right)dv$ in which the fraction $\frac{2}{1+2v}$ is of the required form.

Integrating this gives

$$\int\left(\frac{1}{2}\right)\left(\frac{2}{1+2v}\right)dv$$
$$=\frac{1}{2}\int\left(\frac{2}{1+2v}\right)dv$$
$$=\tfrac{1}{2}\log_e(1+2v)$$

Reverting to equation A,

$$\log_e x+\tfrac{1}{2}\log_e(1+2v) = \text{Constant}$$
$$\therefore\ 2\log_e x+\log_e(1+2v) = \quad ,,$$
$$\therefore\ \log_e(1+2v)x^2 = \quad ,,$$
$$\therefore\ (1+2v)x^2 = \quad ,,$$

But $v=\frac{y}{x}$

$$\therefore\ \left(1+\frac{2y}{x}\right)x^2 = \quad ,,$$

or

$$x^2+2xy = c$$

## LINEAR DIFFERENTIAL EQUATIONS

These are of the general form

$$\frac{dy}{dx}+Py = Q$$

where $P$ and $Q$ are constants or functions of $x$ only. They are called linear equations because $y$ and its derivative are both of the first degree.

This general form may be transformed to

$$dy+(Py-Q)\,dx = 0 \qquad \text{(i)}$$

This is in the recognizable form

$$My+N\,dx = 0$$

where $M = 1$ and $N = Py - Q$. It can therefore be integrated if it is itself an exact equation or can be further transformed to a form which is exact.

In order to find the integrating factor we first assume that there is such a factor $I$ which is a function of $x$ alone. We then set about finding the value of $I$.

First multiply the equation (i) by $I$.

$$I\,dy + I(Py - Q)\,dx = 0$$

If this is exact then the results of differentiating $I$ with respect to $x$

$$\frac{dI}{dx}$$

must be equal to the result of differentiating $I(Py-Q)$ or $IPy - IQ$ with respect to $y$:

$$= IP$$

That is:

$$\frac{dI}{dx} = IP$$

which may be transformed to:

$$\frac{dI}{I} = P\,dx$$

If we now integrate both sides, we obtain:

$$\log_e I = \int P\,dx$$

whence

$$I = e^{\int P\,dx}$$

This, then, is the required integrating factor. Multiplying equation (i) by this factor gives:

$$e^{\int P\,dx}.dy + Py\,e^{\int P\,dx}.dx - Q\,e^{\int P\,dx}.dx = 0 \qquad \text{(ii)}$$

and integrating this gives:

$$y\,e^{\int P\,dx} - \int Q\,e^{\int P\,dx}.dx = c \qquad \text{(iii)}$$

This follows because the integral of each of the first two terms in equation (ii) is $y\,e^{\int P\,dx}$. This integral therefore arises twice on integration but, in accordance with the rule for exact equations, it appears only once in the primitive equation (iii).

Before proceeding further, a note on the integration of $\int Py\, e^{\int P\,dx}.dx$ will be useful. In Chapter 11 it was seen that the derivative of $e^{f(x)}$ is equal to:

$$f'(x)\times e^{f(x)}$$

Consequently if we can recognize a derivative of this form, we may integrate back to $e^{f(x)}$ without much trouble. The expression

$$P\, e^{\int P\,dx}$$

is in fact of this general form. No matter what function of $x$ is represented by $P$, its integral $\int P\,dx$ becomes the exponent of $e$. That is $e^{f(x)} = e^{\int P\,dx}$. But if $f(x)$ is the integral of $P$, then because integration is the inverse of differentiation, so must $P$ be the derivative of $f(x)$. That is $P = f'(x)$. Hence, where

$$f(x) = \int P\,dx$$

so $$P\, e^{\int P\,dx} = f'(x)\times e^{f(x)}$$

and this is in the form of the derivative of $e^{f(x)}$. Consequently by integrating

$$\int P\, e^{\int P\,dx} = e^{\int P\,dx}$$

and $$\int Py\, e^{\int P\,dx} = y\, e^{\int P\,dx}$$

The solution of equation (i) is therefore obtained by first calculating $\int P\,dx$ and then substituting for this in equation (iii). The procedure will become clearer from an actual example. Let the equation to be solved be

$$\frac{dy}{dx}+2xy = 2x^2+1$$

Here $P = 2x$ whence $\int P\,dx = \int 2x\,dx = x^2$

$Q = 2x^2+1$

$e^{\int P\,dx} = e^{x^2}$

(Note that on integrating $\int P\,dx$ we should really show the integral as $x^2+c$ since a constant of integration will normally appear. However, we are merely seeking as simple an integrating factor as we can find. The constant is an arbitrary one and we select the particular case when $c = 0$ and therefore $\int P\,dx$ is equal to $x^2$.) Then, in the form of equation (iii)

$$y\,e^{x^2} - \int (2x^2+1)e^{x^2}.dx = c$$

$$\therefore\; y\,e^{x^2} = \int (2x^2+1)e^{x^2}.dx + c$$

Here we may recognize $(2x^2+1)e^{x^2}$ as the derivative of $x\,e^{x^2}$. This may of course be checked by differentiating. By the product formula, the derivative of $x\,e^{x^2}$ is

$$\begin{aligned} &(x)(2x\,e^{x^2})+(e^{x^2})(1) \\ &= 2x^2\,e^{x^2}+e^{x^2} \\ &= (2x^2+1)e^{x^2} \end{aligned}$$

$$\therefore\; y\,e^{x^2} = x\,e^{x^2}+c$$

and finally,
$$y = x+c\,e^{-x^2}$$

## BERNOULLI EQUATION

This equation is of the form

$$\frac{dy}{dx}+Py = Qy^n \qquad \text{(A)}$$

and is thus not linear. It may however be reduced to a linear form by a change of variable. If we divide this equation by $y^n$ we obtain the new equation

$$y^{-n}.\frac{dy}{dx}+Py^{1-n} = Q \qquad \text{(B)}$$

and if we now substitute in this equation

$$v = y^{1-n}$$

then
$$\frac{dv}{dy} = (1-n)y^{-n}$$

and
$$\frac{dv}{dx} = (1-n)y^{-n}.\frac{dy}{dx}$$

whence $$\frac{dy}{dx} = \frac{dv}{dx} \div (1-n)y^{-n}$$

Equation B then becomes

$$\left[\left(y^{-n}.\frac{dv}{dx}\right) \div (1-n)y^{-n}\right] + Pv = Q$$

Simplifying the expression in brackets and then multiplying the whole equation by $(1-n)$ gives

$$\frac{dv}{dx} + (1-n)Pv = (1-n)Q \qquad \text{(C)}$$

This equation is in the standard linear form where $v$ is the dependent variable. This may be solved for $v$ and we may then find an expression for $y$.

As an example let

$$\frac{dy}{dx} - \frac{y}{2x} = 5x^2y^5$$

Here $$\left.\begin{aligned} P &= -\frac{1}{2x} \\ Q &= 5x^2 \\ n &= 5 \end{aligned}\right\} \text{in equation (A)}$$

and $$v = y^{1-n} = y^{-4}$$

Substituting these values in equation C

$$\frac{dv}{dx} + (-4)\left(-\frac{1}{2x}\right)v = (-4)(5x^2)$$

or $$\frac{dv}{dx} + \frac{2}{x}.v = -20x^2$$

or $$dv + \frac{2}{x}.v\,dx = -20x^2\,dx$$

Multiplying by $x^2$ (the appropriate integrating factor)

$$x^2\,dv + 2xv\,dx = -20x^4\,dx$$

This is an exact equation. Therefore

$$\int x^2\,dv + \int 2xv\,dx = \int -20x^4\,dx$$

$$\therefore\ x^2v = c - 4x^5$$

On integration the term $x^2v$ arises twice but is of course included in the primitive only once (see page 287).

$$\therefore\ v = cx^{-2} - 4x^3$$

But $v = y^{-4}$ so that

$$y = \frac{1}{\sqrt[4]{v}}$$

$$\therefore\ y = \frac{1}{\sqrt[4]{cx^{-2} - 4x^3}}$$

The foregoing examples are intended to indicate the nature of the more common types of ordinary differential equations and the general manner in which they may be tackled. A full treatment is of course not possible here, and the reader who wishes to pursue the subject in considering either more difficult examples or the general theory of equations should consult a specialist work on that aspect of the calculus.

## Chapter 23

# EXPANSION OF SERIES

In Chapter 11 we considered the limiting value, $e$, of the binomial $\left(1+\frac{1}{n}\right)^n$ when $n$ is made indefinitely great. In order to calculate this limit it was first necesssary to transform, or expand, the binomial expression into an infinite series. We could then obtain as near an approximation to the value of $e$ as wc desircd by taking the sum of a sufficient number of terms of the series.

The idea of expanding an expression may be extended to certain other functions. The calculus depends upon the concept of limiting values of infinite series. It is interesting to note that in turn the calculus may be employed to assist in the expansion of functions into convergent series. This has useful applications in the calculations of logarithms, trigonometrical ratios and, among others, the value of $\pi$.

A convergent series is a series the sum of the first $n$ terms of which cannot numerically exceed some finite quantity (i.e. its limit) no matter how great $n$ may be. It is to be distinguished from a divergent series the sum of the first $n$ terms of which may be made numerically greater than any finite quantity by taking $n$ sufficiently great.

Whether or not a series is convergent is not always an easy matter to decide. This however does not concern us here since our object is to demonstrate an application of the calculus to specific instances and not to investigate the general properties of series. During the course of this demonstration we shall find a use for derivatives higher than the second derivative, and it will be assumed that the successive derivatives shown all exist.

The general form of expansion considered here is one which may be equated to a function of the form $f(x+a)$, such that the terms of the expansion involve (i) ascending powers of $a$ and (ii) successive derivatives of $f(x)$.

In describing the function and its derivatives we resort to the alternative symbolism for derivatives. This is merely to simplify the presentation. Thus

$$f(x) = \text{function of } x$$

$$f'(x) = \text{first derivative } \frac{dy}{dx}$$

$$f''(x) = \text{second} \quad ,, \quad \frac{d^2y}{dx^2}$$

and so on.

## TAYLOR'S THEOREM

Taylor's Theorem may now be stated as follows:

$$f(x+a) = f(x)+af'(x)+\frac{a^2}{\underline{|2}}.f''(x)+\frac{a^3}{\underline{|3}}.f'''(x)+ \dots$$

where $f(x)$ is a function of $x$ not involving $a$. It will be seen that the second term of the expansion involves the *first* power of $a$ and the *first* derivative of $f(x)$. The third term involves the *second* power of $a$ and the *second* derivative of $f(x)$. Generally, the $(n+1)$th term involves the $n$th power of $a$ and the $n$th derivative of $f(x)$.

Not every function may be expanded in this way. It may be shown that all the functions discussed here may be so treated, but the proof of this is not within our scope. One important requirement, of course, is that the assumption that a function may be expanded in this way presupposes that all the successive derivatives in fact exist.

To prove Taylor's Theorem we first express the function as a series of increasing powers of $a$ thus:

$$f(x+a) = A_0+A_1a+A_2a^2+A_3a^3+A_4a^4+ \dots$$

where the coefficients $A_0$ etc. are functions of $x$ but do not contain $a$. The problem then is to identify these coefficients.

Since the equation is an identity (that is, the expressions on each side of the equation are always equal for every value of $x$ and $a$) then, if each side is differentiated with respect to $a$ (while $x$ remains constant), their derivatives must also be equal. The first derivative $f'(x+a)$, treating only $a$ as variable, is calculated by first principles term by term.

(i) $A_0$ is constant: its derivative is therefore 0
(ii) The derivative of $A_1a = A_1 \times 1 = A_1$
(iii) The derivative of $A_2a^2 = A_2 \times 2a = 2A_2a$

and so on. That is

$$f'(x+a) = A_1+2A_2a+3A_3a^2+4A_4a^3+\ldots$$

This expansion is in turn differentiated to obtain the second derivative

$$f''(x+a) = 0+2A_2+3.2.A_3a+4.3.A_4a^2+\ldots$$

and again

$$f'''(x+a) = 0+3.2.1.A_3+4.3.2.A_4a+\ldots$$

The theorem applies for all values of $x$ and $a$. Accordingly it holds good for $a = 0$ in particular. When $a = 0$, the function and its derivatives are simplified to

$$f(x) = A_0 \qquad f''(x) = 2.1.A_2$$
$$f'(x) = A_1 \qquad f'''(x) = 3.2.1.A_3$$

and, in general, $f^n(x) = \underline{|n}.A_n$. Rearranging this

$$A_n = \frac{f^n(x)}{\underline{|n}}$$

and the coefficients of the original expansion are identified in terms of $x$. These equivalents may now be substituted in the original expansion. Thus:

$$f(x+a) = A_0+A_1a+A_2a^2+A_3a^3+\ldots$$
$$= f(x)+f'(x)a+\frac{f''(x)}{\underline{|2}}.a^2+\frac{f'''(x)}{\underline{|3}}.a^3+\ldots$$

(or in its usual form)

$$= f(x)+af'(x)+\frac{a^2}{\underline{|2}}.f''(x)+\frac{a^3}{\underline{|3}}.f'''(x)+\ldots$$

## BINOMIAL THEOREM

The expansion of the binomial $(x+a)^n$ can be obtained by elementary algebra methods. It may also be derived from Taylor's Theorem, and may therefore be regarded as a special form of this theorem.

In this instance we have

$$f(x+a) = (x+a)^n$$

When $a = 0$ then

$$f(x) = x^n$$
$$f'(x) = nx^{n-1}$$
$$f''(x) = n(n-1)x^{n-2}$$

and so on.

Substituting these equivalents in Taylor's series will give

$$(x+a)^n = x^n+anx^{n-1}+\frac{a^2}{\underline{|2}}.n(n-1)x^{n-2}+\frac{a^3}{\underline{|3}}.n(n-1)(n-2)x^{n-3}$$

or, in its usual form, where powers of $x$ are shown in descending order and powers of $a$ are shown in ascending order

$$(x+a)^n = x^n+nx^{n-1}a+\frac{n(n-1)}{\underline{|2}}.x^{n-2}a^2 + \frac{n(n-1)(n-2)}{\underline{|3}}.x^{n-3}a^3+\ldots$$

### MACLAURIN'S THEOREM

Taylor's Theorem may be further adapted to simplify calculations. The theorem holds for all values of $x$ and $a$. We first consider the expansion when $x = 0$.

When $x = 0$, then $f(x+a) = f(a)$ and the expansion becomes

$$f(a) = f(0)+af'(0)+\frac{a^2}{\underline{|2}}.f''(0)$$

where, in general, $f^n(0)$ represents the $n$th derivative of $f(x)$ in which we have substituted the value $x = 0$. It could otherwise have been symbolized more clumsily as $f^n_{x=0}(x)$.

This establishes a relationship for all values of $a$. In the special case, when $a = x$, the equation becomes

$$f(x) = f(0)+x\,f'(0)+\frac{x^2}{\underline{|2}}.f''(0)+\ldots$$

and we now have $f(x)$, a function of $x$, expanded into a series involving successive derivatives of $f(x)$ in each of which the value of $x$ is to be taken as zero.

The above method may seem a roundabout way of arriving at

the expansion but it has been included to show the relationship between Maclaurin's series and Taylor's series. The former is a special form of the latter.

Maclaurin's series may be obtained more simply from the basic assumption that $f(x)$ may be expanded to

$$f(x) = A_1 + A_2x + A_3x^2 + A_4x^3 + \ldots$$

Taking the first few derivatives gives

$$\begin{aligned} f'(x) &= A_2 + 2A_3x + 3A_4x^2 + \ldots \\ f''(x) &= \qquad 2A_3 + 3.2.A_4x + \ldots \\ f'''(x) &= \qquad\qquad 3.2.A_4 + \ldots \end{aligned}$$

When $x = 0$, each of these expansions reduces to its first term

$$\begin{aligned} f(0) &= A_1 \\ f'(0) &= A_2 \\ f''(0) &= 2A_3 \qquad \text{and so on.} \end{aligned}$$

Substituting these in the expansion of $f(x)$ gives Maclaurin's series as before.

This series has useful applications notably in connection with the calculation of logarithms and the values of trigonometrical ratios.

## LOGARITHMIC AND EXPONENTIAL SERIES

The logarithm (to base $e$) of a number may be calculated as follows: Let the number be $1+x$ so that its logarithm $\log_e(1+x)$ is in the form of a function which may be expanded by Maclaurin's theorem.

Thus $f(x) = \log_e(1+x)$. When $x = 0$, this becomes $f(0) = \log_e 1$ which is equal to zero. The procedure then is:

$$\begin{aligned} f(x) &= \log_e(1+x) & \text{so} \quad f(0) &= \log_e 1 &&= 0 \\ f'(x) &= \frac{1}{1+x} & \text{,,} \quad f'(0) &= \frac{1}{1} &&= 1 \\ f''(x) &= -\frac{1}{(1+x)^2} & \text{,,} \quad f''(0) &= -\frac{1}{1} &&= -1 \text{ or } -\underline{|1} \\ f'''(x) &= \frac{1.2}{(1+x)^3} & \text{,,} \quad f'''(0) &= \frac{1.2}{1} &&= 2 \text{ or } \underline{|2} \end{aligned}$$

$$f''''(x) = -\frac{1.2.3}{(1+x)^4} \qquad \text{so } f''''(0) = -\frac{1.2.3}{1} = -6 \quad \text{or} \quad -\underline{|3}$$

$$f^n(x) = \frac{(-1)^{n-1}.\underline{|n-1}}{(1+x)^n} \qquad ,, \; f^n(0) = (-1)^{n-1}.\underline{|n-1}$$

These equivalents may now be substituted in Maclaurin's series and

$$f(x) = f(0)+x\,f'(0)+\frac{x^2}{\underline{|2}}.f''(0)+\ldots$$

becomes $\log_e(1+x) = 0+x.1-\frac{x^2}{\underline{|2}}.1+\frac{x^3}{\underline{|3}}.2$

$$= x-\frac{x^2}{2}+\frac{x^3}{3}-\frac{x^4}{4}+\ldots$$

the consecutive terms being positive and negative alternately. The logarithm of $1+x$ may therefore be calculated to any desired degree of accuracy by taking a sufficient number of terms. In practice this series is of use only when $x$ is relatively small. It is however possible to deduce other series from it which simplify the calculation of logarithmic tables.* The logarithms are of course to base $e$.

In Chapter 11 it was shown that the function $e^x$ could be expanded to

$$e^x = 1+x+\frac{x^2}{2}+\frac{x^3}{3}+\frac{x^4}{4}+\ldots$$

This result can now be seen to follow directly from Maclaurin's theorem. If $f(x) = e^x$ then

$$f(x) = e^x \qquad \text{so } f(0) = e^0 = 1$$

The first and every other derivative of $e^x$ is also equal to $e^x$ (see Chapter 11). Consequently $f'(0)$ and all successive derivatives of this form are each equal to 1, and substituting this constant value in the series

$$f(x) = f(0)+x\,f'(0)+\frac{x^2}{\underline{|2}}.f''(0)+\frac{x^3}{3}.f'''(0)+\ldots$$

produces the series for $e^x$ given above.

* See Hall and Knight, *Higher Algebra* (1957), p. 192.

These may be derived as follows:

Let

$$f(x) = \sin x \text{ (where } x \text{ is measured in radians)}$$

then

$$f(x) = \sin x \qquad \text{so } f(0) = \sin 0 = 0$$

$$f'(x) = \cos x = \sin\left(x+\frac{\pi}{2}\right) \qquad \text{so } f'(0) = \sin\frac{\pi}{2} = 1$$

$$f''(x) = -\sin x \qquad \text{so } f''(0) = -\sin 0 = 0$$

$$f'''(x) = -\cos x = \sin\left(x+\frac{3\pi}{2}\right) \qquad \text{so } f'''(0) = \sin\frac{3\pi}{2} = -1$$

and, generally,

$$f^n(x) = \sin\left(x+\frac{n\pi}{2}\right) \qquad \text{so } f^n(0) = \sin\frac{n\pi}{2}$$

Substituting in Maclaurin's theorem

$$f(x) = \sin x = 0+x.1+\frac{x^2}{2}.0-\frac{x^3}{\underline{|3}}.1+\frac{x^4}{\underline{|4}}.0+\frac{x^5}{\underline{|5}}.1$$

$$= x-\frac{x^3}{\underline{|3}}+\frac{x^5}{\underline{|5}}-\frac{x^7}{\underline{|7}}+\ldots$$

The successive derivatives are in turn 1, 0, $-1$, 0 and this rotation continues through all derivatives. When these are substituted in the series, every other term of the series is equated to zero (and thus disappears from the final series for $\sin x$) and the remaining terms are alternately positive and negative.

If $x = 1$ (that is, in radians), then we may calculate the value of the sine of a radian to as great a degree of accuracy as we desire by adding sufficient terms of the series.

The series for $\cos x$ may be derived in very much the same way

$$f(x) = \cos x = \sin\left(x+\frac{\pi}{2}\right) \qquad \text{so } f(0) = \sin\frac{\pi}{2} = 1$$

$$f'(x) = -\sin x \qquad \text{so } f'(0) = 0$$

$$f''(x) = -\cos x = \sin\left(x+\frac{3\pi}{2}\right) \qquad \text{so } f''(0) = -1$$

and so on.

The rotation of derivative values is now in the order 0, $-1$, 0, 1 and, on substitution in Maclaurin's series, we obtain

$$\cos x = 1-\frac{x^2}{\underline{|2}}+\frac{x^4}{\underline{|4}}-\frac{x^6}{\underline{|6}}+\dots$$

This series for cos $x$ could of course be obtained by differentiating the series for sin $x$ (since the derivative of sin $x$ is cos $x$). That is

$$\sin x = x-\frac{x^3}{\underline{|3}}+\frac{x^5}{\underline{|5}}-\frac{x^7}{\underline{|7}}+\dots$$

$$\therefore \cos x = 1-\frac{3x^2}{\underline{|3}}+\frac{5x^4}{\underline{|5}}-\frac{7x^6}{\underline{|7}}+\dots$$

$$= 1-\frac{x^2}{\underline{|2}}+\frac{x^4}{\underline{|4}}-\frac{x^6}{\underline{|6}}+\dots$$

## GREGORY'S SERIES

This series may be simply derived without relying on Maclaurin's Theorem. It may be established by simple division that

$$\frac{1}{1+x^2} = 1-x^2+x^4-x^6+\dots$$

If it now be assumed that $\frac{1}{1+x^2}$ is the derivative of some other function, then the series also represents the same derivative. We can in fact recognize $\frac{1}{1+x^2}$ as the derivative of $\tan^{-1}x$ with respect to $x$. That is

$$\frac{dy}{dx} = \frac{1}{1+x^2} \qquad \text{where } y = \tan^{-1}x$$

whence $$dy = \frac{dx}{1+x^2}$$

and from the above series therefore

$$dy = dx-x^2\,dx+x^4\,dx-x^6\,dx+\dots$$

If we integrate each side of this equation separately with respect to $x$ then the two integrated functions thus obtained must still be equal to each other. That is

$$\int \frac{dx}{1+x^2} = \int dx - \int x^2\,dx + \int x^4\,dx - \int x^6\,dx$$

or $$\tan^{-1}x = x - \frac{x^3}{3} + \frac{x^5}{5} - \frac{x^7}{7} + \ldots$$

This series can be used to calculate the value of $\pi$ but it converges slowly and in practice it is simpler to use a series which converges more rapidly. Nevertheless it does have this property by virtue of the fact that $\tan^{-1} 1 = \frac{\pi}{4}$. If therefore $x = 1$ is substituted in the series

$$\tan^{-1} 1 = \frac{\pi}{4} = 1 - \tfrac{1}{3} + \tfrac{1}{5} - \tfrac{1}{7} + \ldots$$

so that $$\pi = 4(1 - \tfrac{1}{3} + \tfrac{1}{5} - \tfrac{1}{7} + \ldots)$$

A final warning reminder may not come amiss. We have demonstrated *how* certain functions may be expanded. Nothing in this chapter however constitutes or contributes to any proof that the functions considered *can* be expanded or that the series are convergent. We have acted on the knowledge that these proofs have been established although not reproduced here.

## Chapter 24

# FALLACIES

The novelty of the ideas on which the calculus is based and the unfamiliarity of the notation often lead the unwary into the paths of fallacy and error. Fallacies arise out of faulty reasoning, and errors arise out of a misunderstanding either of the notation or of the basic principles.

A short collection of elementary examples now follows, experience having shown that the detective approach required in spotting the fallacy often helps towards a firmer grip on the correct principles and is a useful spur for encouraging revision.

The reader is invited to sort this little lot out. Some are no more than howlers but they have all been perpetrated at one time or another. They are not arranged in any particular order and are first presented without comment. Explanatory notes and references follow at the end of the chapter.

*First fallacy*

*To calculate the square root of* $-\frac{1}{3}$

Let $\qquad y = 3x^2+1 = 0 \qquad (\text{so } x = \sqrt{-\tfrac{1}{3}})$

Then $\qquad \dfrac{dy}{dx} = 6x \qquad = 0 \qquad (\text{so } x = 0)$

$$\therefore\ x = \sqrt{-\tfrac{1}{3}} = 0$$

*Second fallacy*

*To prove that the sum of a series of positive terms may be negative*

Let $\qquad s = 1+2+4+8+16+\ldots$

Then $\qquad 2s = 2+4+8+16+\ldots$

But this is the same series as for $s$ except that the number 1 is missing

$$\therefore\ 2s = s-1$$

and $\qquad s = -1$

*Third fallacy*

*To prove by the Binomial Theorem that* $1 = 2$

The theorem states that

$$(a+b)^n = a^n + na^{n-1}.b + \ldots + b^n \qquad \text{(i)}$$

Let $n = 0$. Then

$$(a+b)^n = (a+b)^0 = 1$$
$$a^n = a^0 = 1$$
$$b^n = b^0 = 1$$

Expansion (i) then becomes

$$1 = 1+0+ \ldots +1$$

since each term (other than the first and the last) on the right-hand side has $n$ as a factor and is therefore zero when $n$ is zero

$$\therefore\ 1 = 2$$

*Fourth fallacy*

*Find the derivative of the function:*

$$y = (x^2+2x)^5$$

Let $\qquad y = u^5 \qquad$ where $u = x^2+2x$

Then $\qquad \dfrac{dy}{du} = 5u^4$

and, reinstating the equivalent of $u$,

$$\frac{dy}{dx} = 5(x^2+2x)^4$$

*Fifth fallacy*

*To prove that the area of any square is zero*

Let the side of the square $= 2a$. Place the square so that its centre coincides with the origin of co-ordinate axes. Then the area of the upper half of the square may be regarded as the area under the linear curve (formed by the upper boundary of the square) between the limits $x = -a$ and $x = a$. The general formula for the area under a curve is $\int y\,dx$. Accordingly the area of half the square is equal to

$$\int_{-a}^{a} y\,dx$$

and the area of the square

$$= 2\int_{-a}^{a} y\,dx$$

In this instance $y$ is constant $= a$

$$\therefore \text{ Area} = 2\int_{-a}^{a} a\,dx$$
$$= 2\Big[ax\Big]_{-a}^{a}$$
$$= 2(a^2 - a^2) = 0$$

*Sixth fallacy*

*Solve the following equation for y (where $a^2$ is a constant)*

$$\frac{d^2y}{dt^2} = 4a^2$$

$$\therefore \frac{dy}{dt} = 2a$$

$$\therefore dy = 2a\,dt$$

$$\therefore y = \int 2a\,dt$$
$$= 2at + c$$

*Seventh fallacy*

*What is the total differential dz of the function:*

$$z = 3x^3 + 4x^2y + y^2$$

Differentiate first with respect to $x$ in order to obtain one partial differential:

$$dz_x = 9x^2 + 8xy$$

Next differentiate with respect to $y$:

$$dz_y = 4x^2 + 2y$$

Total differential is the sum of the partial differentials:

$$dz = 9x^2 + 8xy + 4x^2 + 2y$$
$$= 13x^2 + 8xy + 2y$$

*Eighth fallacy*

*What is the integral of* $\int (x^3+2x^2)\,dx$?

$$\int (x^3+2x^2)\,dx = x^2 \int (x+2)\,dx$$

$$= x^2\left(\frac{x^2}{2}+2x+c\right)$$

$$= \frac{x^2(x^2+4x+c)}{2}$$

*Ninth fallacy*

*The surface area of a sphere is given in textbooks as* $4\pi r^2$. *Prove that this is incorrect!*

A hemisphere may be regarded as consisting of indefinitely small circular discs of width $dx$. The surface area of the hemisphere is therefore equal to the sum of the areas of the circular surfaces of these discs.

The circular surface of each disc may be regarded as an unwound rectangular strip of width $dx$ and length $2\pi y$. The total of all these surfaces is $\int_0^r 2\pi y\,dx$ (if the radius of the hemisphere is $r$) and this is also the surface area of the hemisphere. Then if $s$ = surface area of the sphere,

$$\frac{s}{2} = \int_0^r 2\pi y\,dx$$

$$= 2\pi \int_0^r y\,dx$$

$$\therefore s = 4\pi \int_0^r y\,dx$$

$$= 4\pi \int_0^r \sqrt{r^2-x^2}\,dx \qquad \text{(since } y = \sqrt{r^2-x^2}\text{)}$$

$$= 4\pi\left(\frac{\pi r^2}{4}\right) \qquad \text{(see page 193)}$$

$$= (\pi r)^2 \qquad \text{(i.e. } \textit{not } 4\pi r^2\text{)}$$

*Tenth fallacy*

*Prove that* $\int \frac{a}{x}.dx$ *is indeterminate for any value of a*

$$\int \frac{a}{x}.dx = \int ax^{-1}\,dx$$

$$= \frac{ax^0}{0} = \frac{a}{0}$$

and this is indeterminate.

*Eleventh fallacy*

*Prove that the hypotenuse of a right-angled triangle is equal to the sum of the other two sides*

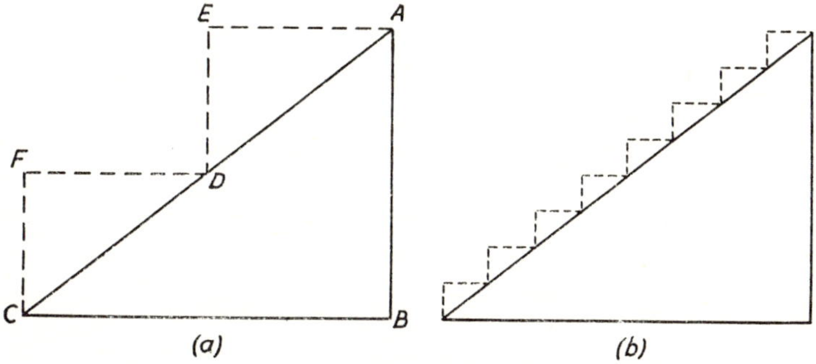


FIG. 68

Fig. 68*a* shows a right-angled triangle ABC on the hypotenuse of which two steps have been drawn so that CD = DA. The sum of the two treads FD and EA is equal to BC. The sum of the two risers FC and ED is equal to AB. Therefore the sum of the four lines drawn to represent the steps is equal to the sum of AB and BC.

Fig. 68*b* shows the same triangle with eight steps formed so that the hypotenuse is divided into eight equal parts. The division of the hypotenuse may be carried on indefinitely; for each division there may be erected a step and the total length of all the lines drawn to represent the treads and risers of all the steps will always be equal to AB+BC.

As the steps grow smaller they approach closer and closer to the hypotenuse. Eventually they become so small that the flight of steps is indistinguishable from the hypotenuse. In the limit therefore the hypotenuse is equal to the sum of all the lines representing the steps. But this sum has already been shown to be equal to AB+BC.

Therefore the hypotenuse CA is equal to the sum of the other two sides AB and BC.

*Twelfth fallacy*

*For what value of x between 1 and 3 is this function a maximum?*

$$\underline{y = x^3 - 6x^2 + 11x - 6}$$

Firstly, $$\frac{dy}{dx} = 3x^2 - 12x + 11$$

Equate this derivative to zero (for a maximum) and solve the quadratic, giving

$$3x^2 - 12x + 11 = 0$$

$$\therefore x = 1{\cdot}42 \quad \textit{or} \quad 2{\cdot}58 \text{ (approx.)}$$

Since 2·58 is the greater, then this is the value of $x$ for which the function has a maximum value.

*Thirteenth fallacy*

*What is the effect on the value of* $y = x^3 + x^2 + x$ *by changing the value of x from* 10 *to* 20*?*

$$y = x^3 + x^2 + x$$

Using differentials

$$dy = (3x^2 + 2x + 1)\, dx$$

In this equation $x = 10$ and $dx$, the differential and/or the increment of $x = 20 - 10 = 10$. Then

$$dy = (300 + 20 + 1)10 = 3210$$

The differential $dy$ is approximately equal to $\Delta y$, the increment in $y$. Hence the effect of changing the value of $x$ from 10 to 20 is to increase the value of $y$ by approximately 3210.

*First*

This fallacy confuses two different values of $x$. The function $y = 0$ only when $x = \sqrt{-\frac{1}{3}}$. The derivative $\frac{dy}{dx} = 0$ only when $x = 0$. That is, the two quantities do not each equal zero for the same value of $x$, and they cannot be associated in the way attempted.

The derivative depends upon $x$ only; it does not equal zero merely because $y$ is zero. It should also be noted that in attempting to equate $3x^2+1$ to zero without restricting this equation to just one value of $x$, we are guilty of misrepresenting the variable $x$ as a constant.

*Second*

Sums of infinite series which are divergent cannot be treated in this way. Neither of the sums has a limiting value.

Series which are convergent may be treated in the way shown. If

$$s = 1+\tfrac{1}{2}+\tfrac{1}{4}+\tfrac{1}{8}+\ldots$$

$$2s = 2+1+\tfrac{1}{2}+\tfrac{1}{4}+\ldots$$

Then $$2s = s+2$$

and $$s = 2$$

which value is in fact the limiting value of $s$.

*Third*

The binomial theorem for positive integers does not hold for $n = 0$.

It is of interest to note that the expansion of a binomial has a finite number of terms only when $n$ is a positive integer. If $n$ is fractional or negative the expansion has an infinite number of terms, and every term except the first has $n$ as a factor.

*Fourth*

The fallacy here is in assuming that $\frac{dy}{du}$ becomes $\frac{dy}{dx}$ when $(x^2+2x)$

is reinstated in place of $u$. In fact

$$\frac{dy}{dx}=\frac{dy}{du}\times\frac{du}{dx}=5u^4(2x+2)$$

$$=5(x^2+2x)^4(2x+2)$$

$$=10(x+1)(x^2+2x)^4$$

*Fifth*

Here the algebra is wrongly shown and ignores the fact that, relative to the $x$ axis, 'one' half of the square has a positive area while the area of the other half is negative. The easiest way to find the area of the square is to calculate the area of the upper right quarter and then multiply by 4:

$$=4\int_0^a y\,dx=4\int_0^a a\,dx$$

$$=4\Big[ax\Big]_0^a=4a^2$$

and this is equal to the square of one of the sides.

*Sixth*

This is a real howler. The symbol $\frac{d^2y}{dt^2}$ which represents the second derivative has been mistaken for the square of the first derivative: $\left(\frac{dy}{dt}\right)^2$.

Correct result:

$$\frac{d^2y}{dt^2}=4a^2$$

$$\therefore\ \frac{dy}{dt}=4a^2t+c_1$$

$$\therefore\ y=2a^2t^2+c_1t+c_2$$

*Seventh*

The error arises in the form of the partial differentials. The partial differential with respect to $x$ is not $9x^2+8xy$. It is $(9x^2+8xy)\,dx$. Similarly the partial differential with respect to $y$ is $(4x^2+2y)\,dy$. The total differential is therefore:

$$dz=(9x^2+8xy)\,dx+(4x^2+2y)\,dy$$

*Eighth*

Although it is permissible to express $(x^3+2x^2)$ as $x^2(x+2)$, nevertheless $x^2$ cannot be transferred outside the integral sign. Only constants may be transferred in this way. Correct method:

$$\int (x^3+2x^2)\,dx = \int x^3\,dx+2\int x^2\,dx$$

$$= \frac{x^4}{4}+\frac{2x^3}{3}+c_1$$

or

$$= \frac{3x^4+8x^3+c}{12}$$

*Ninth*

The surface area of a solid of revolution must not be regarded as composed of the curved surfaces of cylindrical discs of width *dx*. It is to be regarded as composed of the areas of the curved surfaces of cone frustums.*

*Tenth*

The normal rule† for the integration of $\int x^n\,dx$ does not hold for $n = -1$.

$$\int x^{-1}\,dx = \log_e x+c$$

$$\therefore \int ax^{-1}\,dx = a\int x^{-1}\,dx = a\log_e x+c$$

*Eleventh*

This is an optical illusion. The fallacious argument here is that the hypotenuse = AB+BC *because* the total length of all the lines representing the steps = AB+BC. It is indeed this very fact that prevents the hypotenuse from ever being identical with the lines representing the steps! Before we may treat the length of the hypotenuse as the limiting value of the sum of the lengths of the other two sides, it must first be shown that this sum does in fact approach closer and closer to the length of the hypotenuse as the steps are made smaller. This however is not so. No matter what size we make the steps, the sum of the lines representing

* See page 218.    † See page 146.

them is a constant value = AB+BC. This sum does not approach the length of the hypotenuse at all.

If the hypotenuse be regarded as the curve of a function with C placed at the origin of co-ordinate axes, then to suppose that the dotted lines ever become identical with the hypotenuse is tantamount to supposing that the area under the curve (i.e. the area of triangle ABC) may be regarded as consisting of indefinitely small rectangles. In Chapter 9 however it was made clear that this is not so, and that the area must be regarded as consisting of trapeziums.

*Twelfth*

The calculations are correct. The final inference is not.* The value $x = 2{\cdot}58$ in fact makes $y$ a minimum. $y$ is a maximum when $x = 1{\cdot}42$.

*Thirteenth*

The differential $dy$ is approximately equal to the increment $\Delta y$ only when $\Delta x$ is very small† relative to $x$. Increments are always regarded as being very small and if $x = 10$ then the addition of $\Delta x = 10$ is hardly to be regarded as an increment in the true sense.

The use of differentials is therefore inappropriate since $\Delta x$ is not small relative to $x$. When

$$x = 10 \qquad y = 1110$$
$$x = 20 \qquad y = 8420$$

The effect of changing the value of $x$ from 10 to 20 is therefore to increase the value of $y$ by 7310.

* See page 83. † See page 96.

*Appendix I*

# PARTIAL FRACTIONS

The method of splitting a fraction into partial fractions (the denominators of which involve $x$ but no higher power of $x$) is, in general, as follows:

Let the fraction to be split be $\dfrac{4x+5}{x^2+x-2}$

It is assumed that this represents the sum of two fractions so that:

$$\frac{A}{Y}+\frac{B}{Z}=\frac{4x+5}{x^2+x-2}$$

The rule for the addition of fractions is:

(i) Multiply the denominators of the fractions together to obtain the denominator of the total fraction; thus,

$$YZ = x^2+x-2$$

(ii) Multiply the denominator of each fraction by the numerator of the other and add the two products so obtained to form the numerator of the total fraction; thus,

$$AZ+BY = 4x+5$$

To identify $Y$ and $Z$ we now factorize their product:

$$YZ = x^2+x-2 = (x+2)(x-1)$$

Let $Y = x+2$; and $Z = x-1$. This now gives

$$\frac{A}{x+2}+\frac{B}{x-1}=\frac{4x+5}{x^2+x-2}$$

Then, from (ii) above

$$A(x-1)+B(x+2) = 4x+5$$

This equation is true for *all* values of $x$. When $x = 1$ the equation becomes

$$B(3) = 9$$

$$\therefore B = 3$$

When $x = -2$ the equation becomes

$$A(-3) = -3$$

$$\therefore A = 1$$

and substituting in the original equation gives

$$\frac{1}{x+2}+\frac{3}{x-1} = \frac{4x+5}{x^2+x-2}$$

*Appendix II*

# TRIGONOMETRIC RATIOS AND IDENTITIES

*Basic ratios:* In the right-angled triangle ABO (Fig. 69)

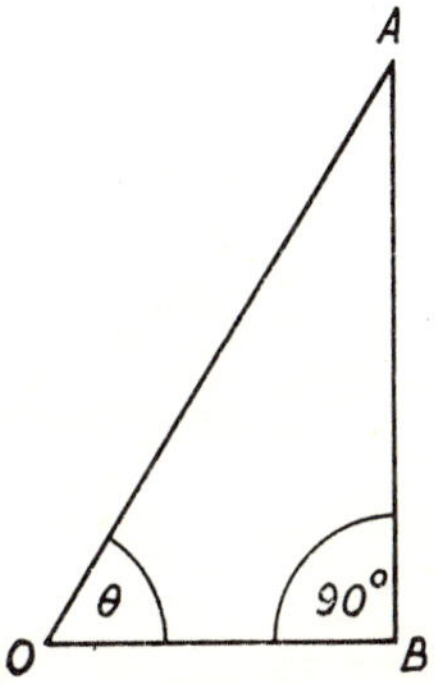


FIG. 69

$$\sin\theta = \frac{AB}{OA} = \frac{\text{opposite side}}{\text{hypotenuse}}$$

$$\cos\theta = \frac{OB}{OA} = \frac{\text{adjacent side}}{\text{hypotenuse}}$$

$$\tan\theta = \frac{AB}{OB} = \frac{\text{opposite side}}{\text{adjacent side}}$$

*Inverse ratios:*

$$\operatorname{cosec}\theta = \frac{1}{\sin\theta} = \frac{OA}{AB}$$

$$\sec\theta = \frac{1}{\cos\theta} = \frac{OA}{OB}$$

$$\cot\theta = \frac{1}{\tan\theta} = \frac{OB}{AB}$$

*Relation between ratios* (angles of any magnitude; explanations are for acute angles only)

$$\tan\theta = \frac{AB}{OB} = \frac{AB}{OA} \div \frac{OB}{OA} = \frac{\sin\theta}{\cos\theta}$$

$$\left.\begin{aligned} \sin^2\theta+\cos^2\theta &= 1 \\ \sin\theta &= \sqrt{1-\cos^2\theta} \\ \cos\theta &= \sqrt{1-\sin^2\theta} \end{aligned}\right\}$$

since $\left(\frac{AB}{OA}\right)^2+\left(\frac{OB}{OA}\right)^2 = \frac{(AB)^2+(OB)^2}{(OA)^2}$

and, by Pythagoras theorem, the denominator and numerator of the final fraction are equal to each other

$$\left.\begin{aligned} \tan^2\theta+1 &= \sec^2\theta \\ \tan\theta &= \sqrt{\sec^2\theta-1} \\ \sec\theta &= \sqrt{\tan^2\theta+1} \end{aligned}\right\}$$

since by Pythagoras theorem $\frac{(OA)^2-(AB)^2}{(OB)^2} = 1$

then $\left(\frac{AB}{OB}\right)^2+1 = \left(\frac{OA}{OB}\right)^2$

$$\left.\begin{aligned} \sin\left(\theta+\frac{\pi}{2}\right) &= \cos\theta \\ \cos\left(\theta+\frac{\pi}{2}\right) &= -\sin\theta \\ \tan\left(\theta+\frac{\pi}{2}\right) &= -\cot\theta \end{aligned}\right\}$$

where $\frac{\pi}{2}$ radians $= 90°$

*Ratios of compound, multiple and sub-multiple angles:*

(Proofs of the following may be found in any elementary textbook on trigonometry. Readers of *Teach Yourself Trigonometry* (E.U.P.) will find the proofs in Chapter VI of that book.)

$$\sin 2\theta = 2\sin\theta\cos\theta \quad (\textit{and } \sinh 2\theta = 2\sinh\theta.\cosh\theta.)$$

$$\sin M-\sin P = 2\cos\left(\frac{M+P}{2}\right)\sin\left(\frac{M-P}{2}\right)$$

$$\cos M - \cos P = -2 \sin\left(\frac{M+P}{2}\right) \sin\left(\frac{M-P}{2}\right)$$

$$\cos^2 \frac{\theta}{2} = \tfrac{1}{2}(1+\cos\theta)$$

$$\sin^2 \frac{\theta}{2} = \tfrac{1}{2}(1-\cos\theta)$$

$$\cos^2\theta = \tfrac{1}{2}(1+\cos 2\theta) \qquad (\textit{and } \cosh^2\theta = \tfrac{1}{2}(1+\cosh 2\theta))$$

$$\sin^2\theta = \tfrac{1}{2}(1-\cos 2\theta)$$

# *Appendix III*

# STANDARD INTEGRALS

$$\int x^n \, dx = \frac{x^{n+1}}{n+1} \quad \text{(except when } n = -1)$$

$$\int \frac{dx}{x} = \log x$$

$$\int \sin x \, dx = -\cos x$$

$$\int \cos x \, dx = \sin x$$

$$\int \tan x \, dx = -\log \cos x \quad \text{or} \quad \log \sec x$$

$$\int \cot x \, dx = \log \sin x$$

$$\int \sin ax \, dx = -\frac{1}{a} \cos ax$$

$$\int \cos ax \, dx = \frac{1}{a} \sin ax$$

$$\int \tan ax \, dx = \frac{1}{a} \log \sec ax$$

$$\int \sec^2 x \, dx = \tan x$$

$$\int \operatorname{cosec}^2 x \, dx = -\cot x$$

$$\int \sec x . \tan x \, dx = \sec x$$

$$\int \operatorname{cosec} x . \cot x \, dx = -\operatorname{cosec} x$$

$$\int \sinh x \, dx = \cosh x$$

$$\int \cosh x\,dx = \sinh x$$

$$\int \tanh x\,dx = \log \cosh x$$

$$\int \coth x\,dx = \log \sinh x$$

$$\int \operatorname{sech}^2 x\,dx = \tanh x$$

$$\int e^x\,dx = e^x$$

$$\int a^x\,dx = a^x \div \log a$$

$$\int \sqrt{a^2+x^2}\,dx = \tfrac{1}{2}x\sqrt{a^2+x^2}+\tfrac{1}{2}a^2 \log (x+\sqrt{a^2+x^2})$$

$$\int \sqrt{a^2-x^2}\,dx = \tfrac{1}{2}x\sqrt{a^2-x^2}+\tfrac{1}{2}a^2 \sin^{-1}\frac{x}{a}$$

$$\int \sqrt{x^2-a^2}\,dx = \tfrac{1}{2}x\sqrt{x^2-a^2}-\tfrac{1}{2}a^2 \log (x+\sqrt{x^2-a^2})$$

$$\int \frac{dx}{a^2+x^2} = \frac{1}{a}\tan^{-1}\frac{x}{a} \quad \text{or} \quad -\frac{1}{a}\cot^{-1}\frac{x}{a}$$

$$\int \frac{dx}{a^2-x^2} = \frac{1}{a}\tanh^{-1}\frac{x}{a} \quad \text{or} \quad \frac{1}{2a}\log\frac{a+x}{a-x}$$

$$\int \frac{dx}{x^2-a^2} = -\frac{1}{a}\coth^{-1}\frac{x}{a} \quad \text{or} \quad \frac{1}{2a}\log\frac{x-a}{x+a}$$

$$\int \frac{dx}{\sqrt{x^2+a^2}} = \sinh^{-1}\frac{x}{a} \quad \text{or} \quad \log (x+\sqrt{x^2+a^2})$$

$$\int \frac{dx}{\sqrt{x^2-a^2}} = \cosh^{-1}\frac{x}{a} \quad \text{or} \quad \log (x+\sqrt{x^2-a^2})$$

$$\int \frac{dx}{\sqrt{a^2-x^2}} = \sin^{-1}\frac{x}{a} \quad \text{or} \quad -\cos^{-1}\frac{x}{a}$$

$$\int \frac{dx}{x\sqrt{x^2-a^2}} = \frac{1}{a}\sec^{-1}\frac{x}{a} \quad \text{or} \quad -\frac{1}{a}\operatorname{cosec}^{-1}\frac{x}{a}$$

$$\int \frac{dx}{x\sqrt{a^2-x^2}} = -\frac{1}{a}\operatorname{sech}^{-1}\frac{x}{a} \quad \text{or} \quad -\frac{1}{a}\log\frac{a+\sqrt{a^2-x^2}}{x}$$

$$\int \frac{dx}{x\sqrt{a^2+x^2}} = -\frac{1}{a}\operatorname{cosech}^{-1}\frac{x}{a} \quad \text{or} \quad -\frac{1}{a}\log\frac{a+\sqrt{a^2+x^2}}{x}$$

## NOTES

The constant of integration has been omitted from each of these integrals. All logarithms shown are to base *e*.

*Appendix IV*

# LOGARITHMIC EQUIVALENTS OF INVERSE HYPERBOLIC FUNCTIONS

The logarithmic equivalent for $\sinh^{-1}\frac{x}{a}$ may be established as follows:

Let $$y = \sinh^{-1}\frac{x}{a}$$

then $$\frac{x}{a} = \sinh y$$

but $$\cosh^2 y = 1+\sinh^2 y$$

$$= 1+\left(\frac{x}{a}\right)^2$$

$$\therefore \cosh y = \sqrt{1+\left(\frac{x}{a}\right)^2}$$

so $$\sinh y+\cosh y = \frac{x}{a}+\sqrt{1+\left(\frac{x}{a}\right)^2}$$

but $$\sinh y+\cosh y = e^y$$

$$\therefore e^y = \frac{x}{a}+\sqrt{1+\left(\frac{x}{a}\right)^2}$$

$$\therefore y = \log\left(\frac{x}{a}+\sqrt{1+\left(\frac{x}{a}\right)^2}\right)$$

$$= \log\left(\frac{x}{a}+\sqrt{\frac{a^2+x^2}{a^2}}\right)$$

$$= \log\left(\frac{x+\sqrt{a^2+x^2}}{a}\right)$$

It should be noted that

$$\log \frac{x+\sqrt{x^2+a^2}}{a} = \log (x+\sqrt{x^2+a^2})-\log a$$

Since $\log a$ is constant then, when the above logarithm results from integration, $\log a$ may be considered as forming part of the constant of integration. When standard integrals are listed, the constants of integration are usually omitted for convenience of expression. It is usual to omit $\log a$ in the same way.

Logarithmic equivalents of other inverse functions may be obtained in much the same way. They are:

$$\cosh^{-1}\frac{x}{a} = \pm\log\left(\frac{x+\sqrt{x^2-a^2}}{a}\right)$$

$$\tanh^{-1}\frac{x}{a} = \tfrac{1}{2}\log\frac{a+x}{a-x}$$

*Appendix V*

# LIMITS AND CONTINUOUS FUNCTIONS – DEFINITIONS

*Limits*

> *$L$ is the limit of the function $f(x)$ for $x = a$ if, given any arbitrarily small number $\varepsilon$, another number $\delta$ can be found such that for all values of $x$ differing from $a$ by less than $\delta$, then the difference between $f(x)$ and $L$ is in absolute value less than $\varepsilon$.*

Thus the function $f(x)$ is said to tend to a limit $L$ as $x$ tends to $a$ if we can make the difference between $f(x)$ and $L$, irrespective of sign, as small as we please by taking values of $x$ near to $a$. For any given small number $\varepsilon$ it is always possible to find a value of $x$ sufficiently close to $a$ to make the difference between $f(x)$ and $L$ less than the number $\varepsilon$.

*Continuous functions*

> *The function $f(x)$ is continuous within certain limits of $x$ if, for any value of $x = a$ in this interval and for an arbitrarily small positive number $\varepsilon$, it is possible to find an interval about $a$ such that for all values of $x$ in this interval the difference between $f(x)$ and $f(a)$ is in absolute value less than $\varepsilon$.*

The basic property of a continuous function is that it can be represented by an unbroken mathematical curve. That is, if the function $y = f(x)$ is continuous at all points between $x = a$ and $x = b$, then as $x$ changes from $a$ to $b$, then $y$ takes at least once every value between $f(a)$ and $f(b)$.

(For general discussions of the above concepts, reference should be made to the main text.)

# INDEX

*(References are to proper names and to the more important aspects of the subjects listed. A discussion ranging over more than one page is listed under the first page number only)*


Acceleration, 4, 61, 65, 223
Antiphon, 25
Approximate methods (differentials), 96
Arc, length of, 211
Archimedes, 25, 104
Area
  circle, 25, 191
  cone, surface, 214
  cycloid, 247
  double integration, 266
  ellipse, 200
  frustum, surface, 216
  irregular shapes, 202
  normal curve, 251, 257
  plane, 191
  rectangle, maximum, 84
  sign of, 197
  sphere, surface, 217, 308, 313
  square, 306, 312
  surface, 214, 220
  under curve, 102, 112, 191, 198, 220, 221
  within limits, 118
Aristotle, 24, 27, 251
Arrow paradox, 70

Ball, W. W. R., 9
Barrow, Isaac, 9, 38
Barrow's Triangle, 9, 40
Berkeley, Bishop G., 46, 64
Bernoulli equation, 293
Binomial theorem, 50, 140, 142, 296, 298, 306, 311

Catenary, 185
Centre of curvature, 274, 277
Centre of force, 230
Centre of gravity, 230
  (*see also* Centroid)
Centre of pressure, 246
Centroid, 232, 254
  circle, 239
  hemisphere, 236
  quadrant, 235
  semicircle, 233
Circle
  area, 25, 191
  centroid, 239
  curvature of, 270
  osculating, 272
Circle of curvature, 272
Circular measure, 167
Compound interest, 136
Cone
  surface area, 214
  volume, 209
Constant of integration, 117, 154, 280
Constants
  differentiation, 54
  integration, 115
Continuous variation, 11, 16, 69, 76, 252, 325
Control limits, statistical, 259
Convergent series, 34, 296
Cooling, law of, 143
Co-ordinate geometry, 5, 17, 24, 68
Cosine
  differentiation, 173
  series, 302
Curvature, 6, 269
  at a point, 271
  average, 271
  centre of, 274, 277
  circle of, 272
  constant, 271
  of circle, 270
  radius of, 272, 277
  total, 270
Cycloid, 247

Definite integral, 117, 119
de Moivre, A., 254
Derivative, 41, 260
  curve of, 81, 90
  equating to zero, 76, 79, 89, 305, 311
  exponential function, 144, 148, 292
  geometrical representation, 74, 76, 94
  maximum point, 75, 79
  minimum point, 77, 79
  negative value, 52, 74
  notation, 41, 62, 93
  partial, 262, 285
  reconstructing curve, 109
  relation to differentials, 94
  second, 62, 78
  sign of, 52, 74
  stationary values, 90
  successive, 63, 223, 297
  turning points, 77, 92
Descartes, R., 9, 24
Deviation, standard, 228, 255, 258
Dichotomy paradox, 30
Difference
  differentiation, 61
  integration, 157
Differential coefficient, 41
  (*see also* Derivative)
Differential equations, 260, 265, 279
  Bernoulli, 293
  exact, 284
  general forms, 282
  homogenous, 288
  integrating factor, 282, 289, 291, 294
  linear, 290
  separable variables, 281
Differentials, 93, 284
  approximate methods, 96, 310, 314
  calculation of roots, 98
  notation, 93
  partial, 260, 275, 284, 307, 312
  relation to derivative, 94
  relation to increments, 94, 96, 310, 314
  total, 262, 307, 312
  uses, 96
Differential triangle, 40
Differentiation (*see also* Derivative)
  basic ideas, 36
  constants, 54
  cosine, 173
  difference, 61
  exponential, 143, 148, 292
  function of function, 123
  hyperbolic, 186
  inverse function, 133
  logarithmic, 146
  methods, general, 49
  partial, 260, 275, 284, 307, 312
  partial fractions, 132
  product, 125, 150
  quotient, 127, 150
  roots, 129, 150
  sine, 171
  substitution, 123, 148, 176
  successive, 62, 177, 223, 297
  sum, 58
  tangent, 174
  trigonometric, 171
Discontinuous variable, 13
Dispersion, 255
Distribution
  frequency, 253
  sampling, 259
Divergent series, 34, 296
Double integration, 266

Ellipse, area, 200
Ellipsoid, volume, 209
Errors, law of, 254
Euclid, 26
Exhaustion method, 104
Explicit function, 15
Exponential function, 135, 142
  differentiation, 143, 148, 292
  series, 143, 296, 301
  value of *e*, 141
Exponents and logarithms, 135

Factor
  integrating, 282, 289, 291, 294
  form, 228
Fallacies, 305
  approximations, 310, 314
  area, 306, 312
  binomial, 306, 311
  differentials, 310, 314
  equating to zero, 305, 311
  function of function, 306, 311
  integration, 308, 309, 313
  limits, 309, 313
  maximum value, 310, 314

notation, 307, 312
partial differential, 307, 312
sum of series, 305, 311
surface area, 308, 313
Falling bodies, law of, 10
Fermat, P. de, 9, 24, 85
Fluent, 69
Fluxion, 42, 64, 69
Form factor, 228
Frequency
distribution, 253
polygon, 252
Frustum, surface area, 216
Function, 14, 16, 74, 76
explicit, 15
exponential, 143, 148, 292
function of, 123, 147, 306, 311
growth of, 16, 135
hyperbolic, 184, 323
implicit, 15
inverse, 133, 146, 187
linear, 21, 36
notation, 15
periodic, 167
trigonometric, 167
Fundamental theorem, 120

Galileo (G.), 10, 24
Gauss, C. F., 254
Gradient, 19, 22, 36
Gregory's series, 303
Guldin's theorem, 237, 241

Hemisphere
centroid, 236
volume, 207
Heracleitus, 13
Histograms, 251
Homogenous equations, 288
Hooke's law, 225
Hyperbolic functions, 184
catenary, 185
differentiation, 186
inverse, 187
logarithm equivalents, 323
Osborne's rule, 185
series, 188
substitution uses, 188

Implicit function, 15
Impulse, 225
Increments, 17
notation, 17, 107
relation to differentials, 94, 96, 310, 314
Indefinite integral, 115, 117, 118, 154
Inertia, Moment of, 241, 255
Infinitely great and small, 27
Infinite series (*see* Series)
Infinitesimals, 2, 3, 45, 47
Infinity of numbers, 12, 27
Inflexion, points of, 89, 91
Integral
definite, 117, 119
indefinite, 115, 117, 118, 154
standard, 320
Integrand, 114
Integrating factor, 282, 289, 291, 294
Integration
arc, length, 211
area, plane, 191
surface, 214
under curve, 102, 112, 191, 198, 220, 221
basic ideas, 100
check, 115
constant of, 117, 154, 280
constants, 115
difference, 157
double, 266
hyperbolic, 188
inverse of differentiation, 107, 108, 114, 147, 155
limits, within, 118
logarithmic, 163
methods, general, 111, 153, 308, 309, 313
notation, 155
partial, 265
partial fractions, 161
parts, by, 163
product, 158, 163
quotient, 158
roots, 162
substitution, 162, 188, 234
successive, 121, 160, 280
sum, 156
trigonometric, use of, 176, 234
triple, 268
volumes, 206
Inverse function, 133, 146, 178, 187

Kepler, J., 101, 200
Kinetic energy, 242

Laplace, P. S. de, 7
Leibniz, G. W. von, 7, 9, 46
Limit of polygons, 25, 27
Limiting values (*see* Series and limits)
Limits of integration, 118
Limits, statistical control, 259
Linear differential equation, 290
Linear function, 21
Logarithms, 145
  differentiation, 146
  hyperbolic equivalents, 323
  integration, 163
  series, 300
  substitution uses, 148

Maclaurin's theorem, 299, 301, 303
Maxima and Minima, 72, 130, 310, 314
  derivatives and, 75, 77, 79
  maximum area (rectangle), 84
    product, 86
    volume, 72, 79, 86
  minimum cost, 88
    journey time, 130
  points, 74, 77, 91
  restriction, 72, 81, 83
  two variables, 262
Mean
  arithmetic, 254
  quadratic (R.M.S.), 228, 255
Mean values, 224, 226
Mechanics, 230
Minima (*see* Maxima and Minima)
Moment about axis
  first, 230, 241
  second, 241
Moment of Inertia, 241, 255
Motion, 24, 70

Napier, J., 145
Newton, I., 1, 42, 46, 64, 69, 143, 201
Normal curve, 251, 254, 257
Normal distribution, 254
Normal law of errors, 254
Notation, 7
  definite integral, 119, 120
  derivative, 41, 62, 93
  differential, 93
  function, 15
  increment, 17, 107
  indefinite integral, 113, 117, 118, 155
  limiting value, 41
  partial derivative, 261
  partial differential, 262, 307, 312
  successive derivatives, 62
  successive integrals, 121, 312
  trigonometric, 167

Organic growth, 135
Osborne's rule, 185
Osculating circle, 272

Pappus' theorem, 237, 241
Paradox
  arrow, 70
  dichotomy, 30
Partial differentiation, 260, 275, 284, 307, 312
Partial fractions, 132, 161, 315
Partial integration, 265
Parts, integration by, 163
Pascal, B., 27
Periodic functions, 167
Pi, series, 25, 304
Polar co-ordinates, 196
Polygon, frequency, 252
Polygons, limiting forms, 25, 27
Pressure, liquid, 244
Primitive (*or* prime), 115, 280
Probability, 251, 257
Product rule
  differentiation, 125, 150
  integration, 158, 163
Pythagoras, 7

Quadrant, centroid of, 235
Quadratic mean, 228
Quality control, 256
Quotient rule
  differentiation, 127, 150
  integration, 158

Radian measure, 167
Radius of curvature, 272, 277
Radius vector, 196
Rectangle, maximum area of, 84
Ring (solid), volume of, 237
Root mean square (R.M.S.), 228, 255
Roots
  approximations, 98

differentiation, 129, 150
integration, 162, 176, 234

Sampling distribution, 259
Semicircle, centroid of, 233
Separable variables, 281
Sequences (*see* Series)
Series and limits, 24, 28, 39, 44, 309, 313, 325
convergent, 34, 296
divergent, 34, 296
notation, 41
sums of, 30, 305, 311
Series, expansion of, 296
binomial, 50, 140, 142, 296, 298, 306, 311
exponential, 143, 296, 301
Gregory's, 303
hyperbolic, 188
logarithms, 300
Maclaurin's, 299, 301, 303
pi, 304
Taylor's, 297, 298, 299
trigonometric, 302
Simpson's rule, 2, 205, 211, 225
Sine
differentiation, 171
series, 302
Solids, volume of, 206
Speed (*see also* Stone, falling), 4, 19, 23, 61, 66, 76, 119, 221
Sphere
surface area, 217, 308, 313
volume, 206
Square, area of, 306, 312
Square roots (use of differentials), 98
Standard deviation, 228, 255, 258
Standard integrals, 320
Stationary values, 90
Statistics, 251
Stone, falling (speed and acceleration), 10, 21, 54, 61, 66, 119, 221
Successive derivatives, 63, 223, 297
Successive differentiation, 62, 177
exponential, 144
Successive integration, 121, 160, 280, 312
Sum
differentiation, 58
integration, 156
Surface areas, 214, 220
cone, 214
frustum, 216
sphere, 217, 308, 313

Tangent, differentiation, 174
Tangent to curve, 38
Taylor's theorem, 297, 298, 299
Total differentials, 262, 307, 312
Triangle, differential, 40
Trigonometric functions, 167
cosine, differentiation, 173
inverse, 178
differentiation, 179
notation, 167
quotient rule, 175
ratios and identities, 317
series, 302
sine, differentiation, 171
substitution uses, 176, 234
successive differentiation, 177
tangent, differentiation, 174
Triple integration, 268
Turning points (*see also* Maxima and Minima), 77, 92

Variables, separable, 281
Variance, 255
Variation (*see also* Function), 11, 16, 69, 76, 252, 325
two variables, 260
Vectorial angle, 196
Vector, radius, 196
Velocity (*see* Speed)
Volume, integration of, 206, 268
cone, 209
ellipsoid, 209
hemisphere, 207
irregular solids, 211
maximum, 72, 79, 86
ring, 237
sphere, 206

Work (kinetic energy), 243
Work of variable force, 224

Zeno paradoxes
arrow, 70
dichotomy, 30